Tel. 50181

LEICESTER POLYTECHNIC

TECHNOLOGY LIBRARY

Please return this book on or before the last date
stamped below. Fines will be charged on books
returned after this date.

24.S.
-5. NOV.

-1. MAY 1979

15. JAN 87.

INTERNATIONAL
ATOMIC ENERGY AGENCY

Reference methods
for marine radio-
activity studies II
1975   £6.50
621.039.7:574(26)

INTERNATIONAL ATOMIC ENERGY AGENCY   1975
Reference methods for marine radioactivity
Studies II

REFERENCE METHODS
FOR
MARINE RADIOACTIVITY STUDIES II

The following States are Members of the International Atomic Energy Agency:

AFGHANISTAN	HAITI	PARAGUAY
ALBANIA	HOLY SEE	PERU
ALGERIA	HUNGARY	PHILIPPINES
ARGENTINA	ICELAND	POLAND
AUSTRALIA	INDIA	PORTUGAL
AUSTRIA	INDONESIA	ROMANIA
BANGLADESH	IRAN	SAUDI ARABIA
BELGIUM	IRAQ	SENEGAL
BOLIVIA	IRELAND	SIERRA LEONE
BRAZIL	ISRAEL	SINGAPORE
BULGARIA	ITALY	SOUTH AFRICA
BURMA	IVORY COAST	SPAIN
BYELORUSSIAN SOVIET	JAMAICA	SRI LANKA
SOCIALIST REPUBLIC	JAPAN	SUDAN
CAMBODIA	JORDAN	SWEDEN
CANADA	KENYA	SWITZERLAND
CHILE	KOREA, REPUBLIC OF	SYRIAN ARAB REPUBLIC
COLOMBIA	KUWAIT	THAILAND
COSTA RICA	LEBANON	TUNISIA
CUBA	LIBERIA	TURKEY
CYPRUS	LIBYAN ARAB REPUBLIC	UGANDA
CZECHOSLOVAKIA	LIECHTENSTEIN	UKRAINIAN SOVIET SOCIALIST
DEMOCRATIC PEOPLE'S	LUXEMBOURG	REPUBLIC
REPUBLIC OF KOREA	MADAGASCAR	UNION OF SOVIET SOCIALIST
DENMARK	MALAYSIA	REPUBLICS
DOMINICAN REPUBLIC	MALI	UNITED KINGDOM OF GREAT
ECUADOR	MAURITIUS	BRITAIN AND NORTHERN
EGYPT	MEXICO	IRELAND
EL SALVADOR	MONACO	UNITED REPUBLIC OF
ETHIOPIA	MONGOLIA	CAMEROON
FINLAND	MOROCCO	UNITED STATES OF AMERICA
FRANCE	NETHERLANDS	URUGUAY
GABON	NEW ZEALAND	VENEZUELA
GERMAN DEMOCRATIC REPUBLIC	NIGER	VIET-NAM
GERMANY, FEDERAL REPUBLIC OF	NIGERIA	YUGOSLAVIA
GHANA	NORWAY	ZAIRE
GREECE	PAKISTAN	ZAMBIA
GUATEMALA	PANAMA	

The Agency's Statute was approved on 23 October 1956 by the Conference on the Statute of the IAEA held at United Nations Headquarters, New York; it entered into force on 29 July 1957. The Headquarters of the Agency are situated in Vienna. Its principal objective is "to accelerate and enlarge the contribution of atomic energy to peace, health and prosperity throughout the world".

Printed by the IAEA in Austria
July 1975

TECHNICAL REPORTS SERIES No.169

REFERENCE METHODS
FOR
MARINE RADIOACTIVITY STUDIES II

Sampling techniques and analytical procedures for the determination
of selected radionuclides and their stable counterparts II:
Iodine, ruthenium, silver, zirconium and the transuranic elements

INTERNATIONAL ATOMIC ENERGY AGENCY
VIENNA, 1975

The International Laboratory of Marine Radioactivity operates under a tripartite agreement between the International Atomic Energy Agency, the Government of Monaco and the Oceanographic Institute.

REFERENCE METHODS FOR MARINE RADIOACTIVITY STUDIES II
IAEA, VIENNA, 1975
STI/DOC/10/169
ISBN 92-0-125275-7

FOREWORD

The present technical report complements that published as Technical Reports Series No.118 by the International Atomic Energy Agency in 1970. Since, at that time, information on certain radionuclides of importance in the context of waste releases into the marine environment was not covered by the report, the necessity of further studies was emphasized.

A second panel on the Reference Methods for Marine Radioactivity Studies was convened by the International Atomic Energy Agency from 30 October to 3 November 1972 in Vienna to study the available information on current analytical methods for transuranic elements and radioruthenium, and for the radionuclides of zirconium, silver and iodine. It was intended to assist researchers by presenting critical reviews of possible reference methods for analysis of these radionuclides.

The global energy situation calls for a faster rate of expansion of the nuclear industry. It is therefore imperative that the impact of such an expansion on the environment should be carefully evaluated on a sound scientific basis. It is hoped that this report will contribute to the production of reliable data in radionuclide measurements on the marine environmental samples, thus in turn contributing to safeguarding the quality of human life.

CONTENTS

SUPPORTING PAPERS

I. INTRODUCTION

1. GENERAL SCOPE AND MAIN OBJECTIVES

Under its Statute the International Atomic Energy Agency is entrusted with promulgating standards and in drawing up internationally acceptable regulations to prevent pollution of the sea by radioactive materials in amounts that would adversely affect man and his marine resources [1]. To implement these tasks the Agency has convened several panels of leading experts on the relevant subjects. The reports of these panels have already been published [2 -4].

One of the important aspects of environmental protection against radio-active contamination is the monitoring of radionuclides released into the environment by nuclear operations. Since the monitoring data constitute the basis for preventive measures or regulatory actions against pollution, the quality of the data should be sufficiently accurate to permit the linking of the data with the radiation doses received by man and his environment. On the other hand, extensive studies are still required to understand the behaviour of radionuclides in the marine environment. These studies are imperative for obtaining the basis of predictions of the fate of radionuclides released, as well as for understanding the biogeochemical processes prevailing in the marine environment. Since the sea is considered as both an immediate and an ultimate sink of the radionuclides released into the environment and, at the same time, interconnects national territories, the results of radio-nuclide measurements, particularly on marine environmental samples, are required to be comparable on an international basis. Up to the present this requirement was not always fulfilled. To find a solution for the required improvement of the comparability of measurements, an ad hoc panel was convened in November 1968 to study the possibility of presenting sets of reference analytical methods for radionuclides of strontium, caesium, cerium, cobalt and zinc, together with their corresponding stable counter-parts, including the collection, storage and preparation of samples for chemical and radiochemical analyses. The report of this panel was published in the Technical Report Series [5]. Since the above-mentioned radionuclides were selected not on the basis of their importance as marine radioactive contaminants, but because relatively abundant data existed on them in a marine context, the panel pointed out at that time that further studies would be necessary on other radionuclides of importance when sufficient data became available. The panel also recommended that the IAEA initiate and promote intercalibration of radionuclide measurements on various matrices of marine environmental samples in order to examine the present-day comparability, as well as to achieve better comparability of such measurements. In accordance with these recommendations, the Monaco Laboratory has been organizing such intercalibrations since 1970. The results of the completed intercalibration exercises have been published [6 - 9]. It emerged from these results that the critical reviews on the existing analytical methods for transuranic elements, radionuclides of ruthenium, zirconium, silver and iodine were especially necessary to improve the comparability of these measurements.

A new panel to survey critically the present status of analytical methods for transuranic elements, radioruthenium, radiozirconium, radiosilver and radioiodine was convened in November 1972 in Vienna. The present report contains the results of that survey. The panel felt that it should not attempt to recommend standard methods for the radionuclides in question, but to indicate the advantages and pitfalls in existing analytical methods to assist the readers in judging the suitability of the methods both for monitoring and for oceanographic research. Since a large part of the previous report [5] was devoted to general procedures for collection, storage and preparation of samples, it was felt that the present report should only indicate where significant differences from standard procedures were required because of the chemical nature of the elements being analysed.

2.   WORKING PROCEDURES

To cover as many aspects as possible of the existing problems in the analytical methods for transuranic elements, radionuclides of ruthenium, zirconium, silver and iodine, detailed information either in the form of reviews or papers on specific points were presented by one or more partici-pants of the panel for each radionuclide. These papers are annexed to this report. On the basis of the material submitted, a detailed survey of the available methods was carried out by three working groups. The results of the technical survey and related recommendations are summarized in sections II -VI of the present report, and the recommendations to the IAEA are presented in section VII.

3.   MEMBERS OF THE STUDY GROUP

The study was carried out by an ad hoc panel under the chairmanship of Dr. J.H. Harley, USA. The participants were:

Panel members

Harley, J.H. (Chairman)   Health and Safety Laboratory,
                               United States Atomic Energy Commission,
                               New York, USA

Aarkrog, A.   Health Physics Department,
                               Atomic Energy Commission
                                  Research Establishment,
                               Risφ, Denmark

Bowen, V.T.   Woods Hole Oceanographic Institution,
                               Woods Hole, Massachusetts,
                               USA

Dutton, J.W.R.   Fisheries Radiobiological Laboratory,
                               Lowestoft, Suffolk,
                               United Kingdom

Georgescu, J.	Polytechnical Institute, Bucarest, Romania
Guéguéniat, P.	Centre de la Hague, Herqueville, France
Mansfield, A.	Czechoslovak Atomic Energy Commission, Prague, Czechoslovakia
Pillai, K.C.	Health Physics Division, Bhabha Atomic Research Centre, Trombay, India
Shiozaki, M.	Hydrographic Department, Maritime Safety Agency, Tokyo, Japan
Wlodek, S.	Central Laboratory for Radiological Protection, Warsaw, Poland

Observers

Kautsky, H.	Deutsches Hydrographisches Institut, Hamburg, Federal Republic of Germany
Peeters, E.	Institut Royal des Sciences Naturelles de Belgique, Brussels, Belgium

IAEA secretariat

Fukai, R.	International Laboratory of Marine Radioactivity, Oceanographic Museum, Principality of Monaco
Gorski, L.	IAEA Laboratory, Seibersdorf
Murray, C.N. (Scientific Secretary)	International Laboratory of Marine Radioactivity, Oceanographic Museum, Principality of Monaco
Nishiwaki, Y.	Division of Nuclear Safety and Environmental Protection

Richman, D.   Industrial Applications and Chemistry
                                     Section,
                                     Division of Research and Laboratories

Rothchild, S.   Division of Research and Laboratories
(Scientific Secretary)

Suschny, O.   IAEA Laboratory,
                                     Seibersdorf

Members of working groups were as follows:

(a) Working Group for the survey on the transuranic elements

    Dr. Bowen (Chairman), Dr. Aarkrog, Dr. Mansfield, Dr. Murray,
Dr. Nishiwaki, Mr. Pillai, Dr. Suschny.

(b) Working Group for the survey on radioruthenium

    Dr. Guéguéniat (Chairman), Dr. Fukai, Dr. Georgescu, Dr. Shiozaki

(c) Working Group for the survey on radiozirconium, radiosilver and
radioiodine

    Mr. Dutton (Chairman), Dr. Kautsky, Dr. Peeters, Dr. Rothchild,
Dr. Wlodek.

II. TRANSURANIC ELEMENTS

1. INTRODUCTION

Semantically this category includes all elements of atomic numbers greater than 92. While small amounts of ^{239}Pu and ^{244}Pu, atomic number 94, have been found in nature, for practical purposes all the transuranic elements in the environment are man-made. Our concern with them stems partly from the general scientific interest of their biogeochemistry and from the usefulness of some of them as tracers of geochemical processes. Mostly, however, we are concerned because of the very high radiotoxicity of several of these elements. Many of the transuranics are produced in nuclear technology and either are released to the environment as a result of the conditions of their use or must be disposed of as part of the radioactive waste problem.

Fortunately, of the 12 transuranic elements now known, only neptunium, plutonium, americium, curium and californium represent legitimate environmental concern today. These five are produced in large quantities as byproducts of nuclear electrical power generation and are certain to be disseminated in the environment to a measurable extent. Table I shows, for each of the elements, the nuclides of environmental interest with some of the most relevant of their nuclear properties.

Of the five elements of concern, four have been found in measurable amounts in some marine environments: plutonium and americium have been disseminated over the whole surface of the globe as constituents of fall-out from nuclear tests, and this may be true of curium also, though there are few data; neptunium, as well as the three above, can be found in some local situations, either as result of waste disposal, or from specific accident or test conditions. Californium has not, to our knowledge, been found in marine environmental samples; it is, however, being used in power sources, or radiation sources, under conditions that could result in its accidental release to the sea, and its production as waste from nuclear power is expected to increase greatly with the adoption of breeder technology.

Chemically, those transuranics with which we are concerned belong to the actinide series, characterized, as are the lanthanides, by the peculiarity of satisfying increasing nuclear charge by addition of electrons to inner orbitals, with the result that the elements of the series show great chemical resemblance to each other, and show the same slight decrease in ionic radius with increased atomic number that characterizes the lanthanides. Furthermore, as in the lanthanides, chemical resemblances are even stronger among the heavier actinides; the five elements of our concern occupy about the centre of the actinide series and offer, consequently, severe problems in separation for analysis, even though they also offer, as do the lanthanides, possibilities of differential behaviour in the environment.

In the following sections sample collection, sample storage, sample processing, radioanalytical procedures, and presentation of data are discussed in this order. In general the reader is referred to IAEA Technical Reports Series No. 118 [5] for an extended discussion of problems of sample collection, storage and processing for marine radioactivity studies. Although it is now dated in the sense that many additional publications are available, few of the

TABLE I.   TRANSURANIC NUCLIDES OF INTEREST IN THE MARINE
ENVIRONMENTAL CONTEXT

Atomic No.	Element	Nuclides	Half-life[a]	Specific activity (Ci/g)	Radiation and energy (MeV)[a]
93	Neptunium	237	2.14×10^6	7.07×10^{-4}	α 4.79; 4.77
		239	1.5×10^{-2}	1.0×10^5	β^- 0.33; 0.43
94	Plutonium	238	87.8	17.2	α 5.50; 5.46
		239	2.44×10^4	6.14×10^{-2}	α 5.16; 5.14
		240	6.54×10^3	2.38×10^{-1}	α 5.17; 5.12
		241	15	1.0×10^2	β^- 0.021
95	Americium	241	433	3.45	α 5.49; 5.54
96	Curium	242	0.22	6.83×10^3	α 6.11; 6.07
		243	28	54.1	α 5.78; 5.23
		244	17.9	82.8	α 5.81; 5.76
98	Californium	252	2.63	5.46×10^2	α 6.12; 6.07

[a] Principal energy first; Data from Chart of the Nuclides, Knolls Atomic Power Lab., Eleventh
Edition.

points made are not still pertinent, and even fewer of the problems posed
have yet been answered. Only few instances, where the transuranic elements
may be expected to pose unforeseen problems because of their chemical
peculiarities or of special conditions that attend their introduction to the
marine environment, are addressed in this section.

2.   SAMPLE COLLECTION

2.1. General

It does not now appear that the transuranic elements so far examined
offer any special problems in respect to collection of air, water or biological
samples, except those that are consequent on their low concentrations in
most present-day environments and the resulting need for very large samples.
It had been anticipated that elements of such very strong surface activity,
in seawater, might interact with the walls of water sampling devices and
require special procedures for removal. Experience of Bowen and his co-
workers and of Aarkrog with plutonium and americium analyses of large
volume seawater samples offers no evidence of this; doubtless, however,
investigators should bear in mind this danger, especially in the case of
sampling for the heavier members of the series.

   This may be the place to note another sort of hazard consequent on the
difficulty of monitoring the pure alpha emitters among this series: In the
case of spills or other accidental releases it is a great convenience, and
consequently a great temptation, to guide the collection of samples for
analysis by monitoring any more easily detectable radiations that may be
available. After the Thule release, for instance, sampling and clean-up
for plutonium, the material of greatest hazard, was guided by monitoring
the gamma radiation from the americium-241 [10] with which plutonium is
usually contaminated. Such procedures should be followed only with great
caution, and in the realization that little information is available on how
rapidly or how completely the various actinides separate in nature. It has
recently become evident that specific activity considerations and 'hot atom'
chemistry both play important roles in the mobility of individual actinide
nuclides in soils or in organisms. For instance, ^{238}Pu is much more mobile
than is ^{239}Pu [11-13], an observation that is explained by the tendency of
more of the lower specific-activity nuclides to exist as polymers. In one
area of interest the ratio ^{239}Pu to ^{238}Pu varied from 20 in surface soil to
10 in vegetation to X in rodents feeding on the vegetation [14]. In many
media [15, 16] the solubility of ^{241}Am exceeds that of either ^{238}Pu or ^{239}Pu,
but here it is difficult to distinguish the specific activity effect from the
difference in element chemistry. In complex mixtures one would expect
to see a difference between the behaviour of ^{241}Am deriving from decay of
^{241}Pu and that of ^{241}Am originally introduced, simply because of the kinetic
energy with which the daughter atoms are supplied (hot-atom chemistry);
clear-cut cases of this effect in the environmental chemistry of the transuranics are not known, although it is now well-established as responsible
for differences between ^{238}U and ^{234}U distributions in nature.
   It appears that the transuranics may exist in some marine sediments in
conditions that indicate special sampling precautions. Should they be associated, as they are known to be in many soils (Volchok, personal communication),
preferentially with the finer particle sizes, or, as suggested by Noshkin and
Bowen [17], with organic detritus, then the transuranic content of a sediment
could be altered drastically by resuspension during retrieval of any core or
grab sampler that is not hermetically sealed at the top, or during the
commonly practised siphoning off of the water above the sediment in the core
barrel. The responsibility clearly is on the investigator to ensure that no
such losses occur. More subtly, it should be also anticipated that the fines
or the highly organic fractions of a sediment will be more subject to translocation, downward, along the jaws of a grab sampler or along the barrel
of a coring device; accordingly, special attention should be given to careful
trimming, from the sediment samples retrieved, of the probably contaminated
outer rind. Bowen and his coworkers have evidence that quite frequently,
even with the most careful trimming, the sediment at the very bottom of a
core, within the nose cone, will be contaminated with a few per cent of
surface sediment. The effect is more easily seen in the ^{137}Cs profiles of
the series of fresh-water cores illustrated in Table II, because the concentrations are so much higher, but the same tendency is seen in the ^{239}Pu
curves. Of the 13 cores in this series six show higher ^{137}Cs in the bottom
segment — or the bottom two segments — than in the three preceding segments.
Since this shows no correlation with length of core, it is believed to be produced by transfer of surface sediment downward in the nose cone. Volchok
(private communication) recently observed this effect in a shallow water

TABLE II. CAESIUM-137 IN LAKE ONTARIO 21-cm DIAMETER CORES[a]

Core No.	71-2	71-3	71-4	71-5	72-1	72-2	72-3	72-4	72-5	73-1	73-7	73-13	73-14
Activity in top section	2220±17	170±90	3000±14	1030±8	65±35	940±10	2520±11	13550±30	10600±40	286±7	2300±20	9800±30	22400±40
Mean activity for 3 slices above bottom	-2±9	56±5	2±17	587±40	-2±6	2±5	9±7	10±5	96±20	3±6	94±20	0±4	0±5
Activity in bottom slice	450±7	300±6	1±17	364±5	2±4	2±8	49±7	0±7	43±7	35±3	119±4	16±4	30±4
Total length of core (cm)	14	40	40	40	83	17	57	40	61	62	13	82	52

[a] Activity in dpm per kg dry sediment.

core with a surface layer that was much paler than the underlying sediment;
obvious surface material had been included in the nose-cone contents.
Probably the bottom 3-5 cm of sediment should be discarded; certainly its
content of recently deposited radioactivity should be viewed with grave
suspicion.

As emphasized above, the chemical and physical state of the transuranic
elements in environmental samples can be expected to depend strongly on the
different conditions of their introduction: it cannot be assumed that a sampling
procedure proved to be satisfactory for fall-out-contaminated sediments is
comparably satisfactory for those contaminated by dissolved transuranics in
waste effluents or by oxide particles released in nuclear accidents. Some
data are available referring to plutonium following various modes of introduction, but comparison of the information is difficult because of additional
special features of each case. However, with respect to plutonium at least
one can predict a reasonable progression of environmental behaviours. The
simplest behaviour, that of relatively large particles of plutonium dioxide,
would be seen after an accidental burn-up like that at Thule [10] or after
'safety-shots' (TNT-destructs of experimental nuclear devices). The next
simplest, and that representing (Bowen believes) the end member of the
series of behaviours, would be that of global fall-out, in which plutonium
and americium appear to behave like free ions. In waste effluents, by
contrast, the free ion behaviour appears to be appreciably affected by the
amounts of carrier cations or of organic ligands in the waste; and in close-
in fall-out one seems to be faced by plutonium dioxide particles of a wide
range of sizes, plus complex aggregates and some free ions.

2.2. Accidental release

Most of the plutonium released to the marine environment after the
Thule, Greenland, accident in January 1968 [10] was found in the sediments
as PuO_2 particles. In August 1970 a core sample was collected at the point
of impact. The depth of water was approximately 200 metres. The sediments consisted of clayish mud and the sample was collected with the 21-cm
diameter Woods Hole coring device [18]. The core was 10 cm deep and
Table III shows the activity distribution of the plutonium. Penetration to such
depths in the sediment does not appear unreasonable in an area with a
reasonably rich bottom fauna and with an active seasonal reworking of the
sediment by ice. It is clear that a grab sample in this area would give an

TABLE III. PLUTONIUM-239 IN THULE SEDIMENTS

Layer	mCi/km^2	%	pCi/g dry weight
0 - 3.3 cm	124	62	5.7
3.3 - 6.7 cm	43	22	2.0
6.7 - 10 cm	32	16	1.5
0 - 10 cm	Σ 199	Σ 100	x̄ 3.1

incorrect picture of the sediment load of plutonium and it is likely that the
short core obtained (because of the difficult mechanical properties of the
sediment) has resulted in a considerable underestimate of the inventory of
plutonium in the sediment. So far no data are at hand that relate to the
questions of differential distribution of plutonium and americium or of
association of plutonium preferentially with any sub-population of the
sediments. A new set of core samples was collected in Thule area in
August 1974 (Aarkrog, private communication), which are now being studied
for vertical distribution of ^{239}Pu.

2.3.   Fall-out introduction

     Bowen and co-workers have examined the distributions of fall-out-
delivered transuranic nuclides in both shallow and deep-water marine
sediments. A report of plutonium in the latter class has been published [17].
In the shallow water sediments of Buzzards Bay, Massachusetts, they have
been collecting cores since 1964, using both 7.5 and 21 cm gravity corers
and either direct extrusion or freezing followed by sawing to subdivide the
sediment columns. These studies have shown that carefully taken cores
retain all the plutonium that has been predicted to be delivered to the area;
freezing and sawing cannot be shown, convincingly, to cause significant
vertical translocation along the core; as noted above, with both corers used
nose-cone samples often can be argued to be contaminated by superficial
sediments; plutonium, and probably americium, can be argued to have been
translocated in both directions in these sediments over the 10 years of
sampling. All of this work is now being prepared for publication. Experiments have just been begun by this group to examine Buzzards Bay sediments for preferential associations of plutonium and americium. Very
preliminary data indicate less tendency than was anticipated for plutonium
to be concentrated in the finer sieve-sizes, but no information is yet
available on its relationship to organic matter.
     Plutonium-239 to americium-241 ratios in cores present so far a
confusing picture, in some coastal cores showing significant vertical variations
and in others and in deep ocean cores no variation at all. Since this ratio does
appear to vary systematically in N. Atlantic water columns, and in a way that
shows the two nuclides to be undergoing different sedimentation processes,
it does not seem likely that the growth of ^{241}Am from its parent, ^{241}Pu, can
be used in any simple way for age calculations referring to environmental
samples.

2.4.   Waste effluent

     In the Irish Sea, resulting from Windscale operations, and in Bombay
Harbour measurable amounts of transuranic elements have been released
to coastal marine waters.
     Periodic examination of the bottom sediments in Bombay Harbour and
Bombay Bay has revealed that plutonium is accumulated in the discharge
location and its close neighbourhood. Because of the heavy silt load in the
bay waters and the high distribution factors (K_d) for plutonium, the silt
scavenges the plutonium, which thus is localized near the discharge area.
The distribution factors obtained for plutonium in shore sediments also

indicated that with distance there was less concentration of plutonium in the
shore sediments on either side of the discharge area. A sediment column
of length 11.5 cm collected during low tide from the shore close to the
discharge location showed uneven distribution of plutonium in different
layers. This was to be expected because of differences in discharge rates
and in siltation rates in the bay, and of the resuspension of sediments by
wave action and tidal currents.

Plutonium, americium (and curium) nuclides are present at low concentration in the Windscale discharge and it has been said that these are
immobilized near the discharge point by association with suspensoids and
sediments; data on their concentrations or distributions in Irish Sea sediments are being prepared for publication by the Lowestoft Laboratory.
Such data will have great value in speaking to the question of differential
mobility of these nuclides in marine situations.

3. SAMPLE STORAGE

With respect to the storage of samples it may be expected that the
transuranic elements in biological or sediment samples will offer no special
problems. Organisms preserved in solutions (formaldehyde or alcohol)
must be assumed to have lost material to the solution, which consequently
should also be analysed. Bowen's co-workers (Gschwend, private communication) have, however, found that near-shore sediments can be
exhaustively sieved with ethanol without measurable loss of plutonium.

Sediment samples that have dried out must be treated with care lest
the fines be of higher specific activity than the bulk of the sample.

Water samples do, however, present problems, because of the strong
assocation of the transuranic nuclides with surfaces, which are akin to
those discussed in the previous report [5] concerning the lanthanide nuclides;
the transuranic nuclides seem to be even more difficult to remove from some
surfaces once the association is established. Data are available showing the
association of plutonium with glass, sediment or polyethylene surfaces.

TABLE IV. SORPTION OF PLUTONIUM ONTO GLASS SURFACE

Volume of seawater (ml)	Plutonium added (in effluent) (dpm)	Period of contact	Plutonium sorbed on glass (dpm)	Sorption (%)
100	94.6	3 days	10.35	10.9
100	47.3	3 days	1.87	4.0
100	215	15 months	47	21.9
100[a]	215	15 months	63.4	29.5

[a] Organic matter extracted from sediments added to seawater.

3.1. Sorption on glass

Some experiments showed high sorption of plutonium onto glass surfaces.
Effluents containing plutonium were added to filtered (0.22 μm) seawater from
Bombay Harbour in Pyrex glass containers and allowed to remain in contact
for a certain period. The seawater was subsequently removed and the
plutonium sorbed on the glass surface was estimated by washing the containers
with 4$\underline{N}$ HNO_3, evaporating to dryness and counting the alpha activity. Some
of the results obtained are given in Table IV.

In the last two experiments shown in Table IV it was found that plutonium
adhering to the glass could not be removed by one wash with 4$\underline{N}$ HNO_3. It
required four successive washings with 4$\underline{N}$ HNO_3 to remove completely the
sorbed plutonium on the glass surface.

3.2. Sorption on suspensoids

In the Bombay harbour bay where low-level liquid effluents are released
(after treatment) from a fuel reprocessing plant it was observed that plutonium levels in the water are very low, even in areas close to the discharge
location. The plutonium activity levels in waters were only 10-100 times
the fall-out levels in the oceans. The silt showed high pick-up of plutonium.
It showed distribution factors (K_d) of the order of 10^5, which was the maximum
obtained among other radionuclides like ^{90}Sr, ^{106}Ru, ^{137}Cs, ^{144}Ce etc. in silt.

Clearly, when analyses are to be made for transuranics in water samples
with significant silt loads, either the transuranics must be carefully removed
from the silt surfaces while the sample is still in the original container or
stringent precautions must be taken that all the original suspended matter
is transferred from the container with the water sample.

3.3. Sorption on plastics

Because most seawater samples are stored in polyethylene containers,
a good deal of experience has been accumulated that speaks to the question
of loss of plutonium to container walls and of the desorption of such plutonium.

Aarkrog believes that plutonium is desorbed from the walls of cross-
linked polyetheylene drums by the acidification steps of his normal procedure
[19]. Bowen and his co-workers, who have used both cross-linked and
linear polyethylene containers, have tested their desorption procedure in
several ways: Aliquots of a surface seawater sample, collected in linear
polyethylene, received in one case the ^{236}Pu yield monitor at the time of
collection, and in the other not until the time of desorption before analysis,
12-18 months later; in one aliquot 0.17 ± 0.06 dpm ^{239}Pu was found, and in
the other 0.15 ± 0.05. Evidently no measurable loss of plutonium occurred
on the container walls after desorption as described by Wong et al.(1970) [20].
Aliquots of an inshore seawater sample, used routinely as a quality control
standard, have been analysed at intervals since just after collection (in
November 1972). There has been no systematic change with time in the
^{239}Pu concentration measured, as would be expected if irreversible losses
were taking place to the container walls.

Aliquots of the IAEA seawater sample, SW-A-1, were analysed at once
by the laboratory in Woods Hole, and recently, at least 18 months later,
by Mrs.D. Sutton at the USAEC Health and Safety Laboratory, New York
City. The results, in each case the mean of 4 aliquots measured, were:

WHOI 0.275 ± 0.03 dpm ^{239}Pu per 100 litres

HASL 0.263 ± 0.04 dpm ^{239}Pu per 100 litres

These data, again, give no indication of systematic loss of plutonium during storage. Although no such specific data exist for americium, good desorption for this nuclide also is indicated by the following: two separate stations of large volume water samples, one collected in June 1972, the other in August 1972, were analysed about one year apart. In each case the ^{241}Am:^{239}Pu ratio showed the same increase with depth, from 0.16 in the upper 200 m, to 0.36 - 0.38 at 2000 m or more. Further tests are in progress dealing with americium desorption.

As a part of an early IAEA seawater intercalibration study, a number of different plastic materials were tested as sample containers. These experiments as they concerned plutonium are summarized and discussed by Fukai and Murray [8], and as they concerned a variety of other waste-product nuclides, summarized by Fukai et al. [6] and discussed in detail by Dutton et al. [21]. The latter reference is important to the consideration of americium because the data presented show that removal of ^{144}Ce, once adsorbed on some containers, offers a special problem. The generally more lanthanide-like chemistry of the transplutonium actinides suggests that the cerium-144 experiments may tell us more about how americium — and curium or californium — will behave than do studies with plutonium.

4. SAMPLE PROCESSING

4.1. General

Few special precautions in sample processing appear to be required for the transuranic elements. It must, however, always be borne in mind that the treatment required may depend on the mode of release of the nuclides to be analysed.

Since all the procedures that are in current use give more or less variable chemical yields, it is necessary for accurate work to monitor the yield of each nuclide sought, and since none of the transuranics occur in stable nuclear configuration, these yield monitors must be nuclides of the element sought, which are distinguishable from those found in the environment. This, of course, implies that yields cannot be monitored with nuclides that have been released to the environment studied, and in most cases this presents no problem:

Neptunium is usually sought as either ^{239}Np or as ^{237}Np, and the other can be used as monitor.

Plutonium in the environment is always free of ^{236}Pu, and usually so low in ^{242}Pu that both are available for use as monitors.

Americium is usually sought as ^{241}Am, and ^{243}Am is available for monitoring.

Curium does present a problem in that it may be sought as ^{242}Cm, ^{243}Cm or ^{244}Cm, and none of these is now widely available in good purity to use as monitor. As noted in the section on Radioanalytical Procedures, in at least some methods there is little or no shift in Cm : Am ratio during radiochemical separation. This permits use of ^{243}Am to monitor both ^{241}Am, and any curium nuclide sought.

Californium has not yet been sought in marine samples, and no pertinent data are available. It now appears possible at least that ^{251}Cf and ^{252}Cf may be used in technologies that could result in their release to the environment; it seems that no data have been made available on the occurrence of californium nuclides in power reactor fuel wastes. It appears that some effort should be made to make available high purity ^{248}Cf as a yield monitor and for methods development.

The stipulation is commonly made that after addition of the yield monitor, one or more oxidation-reduction cycles should be gone through, to ensure isotopic equilibrium of monitor with the nuclides sought. This is surely an important precaution in the case of plutonium, and probably of neptunium as well. The difficulty of obtaining the transplutonic elements, in aqueous chloride solutions, in anything but the 3+ state suggests that in their case oxidation-reduction may be illusory, and that the analyst of marine samples should rely rather on very strong acidification of his sample, followed by a prolonged equilibration time. Little data are available that relate directly to this problem; in a recent experiment (Gschwend, private communication) on the distributions of fall-out plutonium among near-shore sediment particles of various sizes or densities, identical yields of added ^{242}Pu tracer and of the fall-out ^{239}Pu (separately measured) were obtained, even though no equilibration procedure had been possible.

In addition to the nuclide yield monitors, any stable element carriers should also be added as early in the procedure as possible. A problem may arise here if the sample is to be analysed both for transuranic nuclides and any of the lanthanide fission products: for ease in the analysis of americium (and doubtless of curium or californium) the total lanthanide content of the solution should be kept as low as possible: the range of 50 mg (Bojanowski et al.[22]) is convenient. Keeping to so little stable carrier might present real difficulty if, for instance, one were following the procedure outlined by Wong et al.[20] for ^{147}Pm. No experiment directed to resolving this difficulty has been made, but it is sure that it will arise, since both ^{147}Pm and ^{151}Sm are tracers of great interest.

Since the treatments demanded for water, biological or sediment samples are so very diverse, it is best to consider the special precautions involved separately. One general statement should be prefaced: Whenever the source of transuranic contamination has been the destruction by fire of a nuclear device, then the samples should always, finally, be put into solution by fusion with sodium carbonate or with sodium bisulphate. Aarkrog, for instance, rigorously follows this latter procedure for his Thule samples, and it should probably be followed for close-in fall-out situations as well. Isotope ratio data seem to indicate that the debris from the SNAP-9A burn-up was so finely divided as to behave like world-wide fall-out, but no critical data from the marine environment are available that show there is no non-leachable residue rich in ^{238}Pu.

4.2. Water samples

In the previous section the reasons have been quoted for believing that several acidification and carrier addition procedures are sufficient to bring all plutonium and americium in large seawater samples into solution and into equilibrium with added monitors. Although these reasons are not wholly satisfactory, they are better than we have for neptunium, curium or californium, where no information exists and even opinion rests only on their resemblance, chemically, to plutonium or americium.

Most procedures where details are specified have followed Sugihara et al. [23] in bubbling tank nitrogen gas through the sample for agitation during desorption. Recently Bowen and his co-workers, following a suggestion by L. Labeyrie, have changed to convective agitation, using low temperature aquarium heaters; it has appeared safer to extend the desorption period, using this method, to 4 - 5 days, but the results are proving entirely satisfactory and there is a significant saving of expense.

4.3. Biological samples

Either wet ashing with nitric, sulphuric and perchloric acid or dry ashing about 450 - 550°C are acceptable treatments for biological material. At Lowestoft (private communication) algal samples after either low-temperature ashing or ashing in a 450°C oven are leached with boiling 4M HCl for analysis of ^{239}Pu and ^{241}Am. Because of the danger of formation of stable complexes of the transuranics it is important that all organic matter be destroyed; in ignition there is danger of forming very insoluble forms, especially of plutonium, if samples are overheated.

Pillai has tested his procedure of wet ashing followed by three extractions of the residue with 8N HNO$_3$ on an algal sample contaminated by waste effluent. The residue still insoluble after this treatment was finally decomposed, and proved to contain only 1.1% of the total plutonium activity.

On the IAEA Fucus sample AG-I-1 [8] and ^{241}Am has been reported in good agreement by three laboratories, two by radiochemistry after wet ashing, the other by non-destructive gamma spectrometry; this can be used to argue that the wet ashing procedure is adequate for ^{241}Am.

Although this is often not done, it is preferable to add carriers and yield monitors to the fresh samples, coarsely minced, before any processing, rather than after either ignition or acid treatments.

4.4. Sediment samples

A good deal of controversy exists over the dissolution of plutonium or other transuranics from sediment samples. It has been noted above that Aarkrog routinely uses complete dissolution by fusion on his sediments from Thule, and in this special case no other course would seem acceptable. In many other cases, however, and especially with fall-out or waste effluent contamination, careful acid leaching appears to be sufficient for good recovery of plutonium or americium.

Pillai has found that leaching coastal sediments with 8N HNO$_3$, repeated three times, was sufficient to remove plutonium quantitatively. In the case of two coastal sediments (one from Buzzards Bay and another from Bombay Harbour Bay) this was checked by decomposing the residue (after extraction with 8N HNO$_3$) with nitric acid, perchloric acid and hydrofluoric acid and estimating plutonium, which was less than detection limits. Although the use of HNO$_3$-HClO$_4$-HF mixtures appears a convenient approach to dissolution of ash, soil or sediments, the recovery from such reactions should always be

TABLE V. RECOVERY OF PLUTONIUM FROM SEDIMENT SAMPLES
BY ACID LEACHING (Pu data in dpm per kg dry)

Sample and acid treatment	$^{239,\,240}$Pu in leach	$^{239,\,240}$Pu in residue	^{242}Pu % in residue	Total $^{239,\,240}$Pu in recovery
Lake sediment				
8M HNO$_3$	12.8 ± 1.3	0.6 ± 0.8	1.4	13.4
8$\underline{M}$ HNO$_3$	12.2 ± 1.1	0.8 ± 0.4	0	13.0
8M HNO$_3$	11.2 ± 1.0	1.2 ± 0.3	0.4	12.4
8$\underline{M}$ HNO$_3$	10.8 ± 1.0	0.7 ± 0.2	0.1	11.5
Lake sediment				
8$\underline{M}$ HNO$_3$ + 12$\underline{M}$ HCl	10.5 ± 0.8	0.3 ± 0.2	4.5	10.8
8$\underline{M}$ HNO$_3$ + 12$\underline{M}$ HCl	10.4 ± 0.9	1.3 ± 0.3	4.1	11.7
8$\underline{M}$ HNO$_3$ + 12$\underline{M}$ HCl	9.8 ± 1.0	1.1 ± 0.3	5.3	10.9
8$\underline{M}$ HNO$_3$ + 12$\underline{M}$ HCl	11.2 ± 0.9	0.6 ± 0.4	1.9	11.8
Buzzards Bay sediment				
8$\underline{M}$ HNO$_3$	52.4 ± 2.2	1.4 ± 0.3	2.9	53.8

carefully monitored. Danger exists that any plutonium oxidized to the 6+
state by anhydrous $HClO_4$ may be lost as PuF_6 (B.P. 62°C).

Bowen and co-workers tested their leaching procedures on both fresh-
and salt-water sediments as follows: 50-gram aliquots of finely ground and
mixed fresh-water sediment and of one core section from Buzzards Bay
were leached, either with 200 ml of 8$\underline{M}$ HNO$_3$ or with 200 ml 8$\underline{M}$ HNO$_3$ +
100 ml 12$\underline{M}$ HCl; about 2 dpm of ^{242}Pu (carefully calibrated) was added as
a yield monitor before the acid leaching. The residues from the leaching
were in turn enriched with about 2 dpm of ^{236}Pu (carefully calibrated) as a
second yield monitor, dried and then fused, in 2.5 g portions, with 15 g
per portion of 1:1 K_2CO_3 and Na_2CO_3; the combined fusions were put into
solution and plutonium separated and prepared for counting. Similarly,
plutonium was separated and prepared for counting from each of the acid
leaches. The results are given in Table V.

Even a single leaching with 8$\underline{M}$ HNO$_3$ was clearly enough to give acceptable recoveries of fall-out ^{239}Pu from these sediments; only in one of the five
cases was the Pu left in the residue more than the uncertainty of the amount
found in the leachate. These data can be taken as further confirmation that
Pillai's procedure of three successive leaches with 8$\underline{M}$ HNO$_3$ would be very
generally acceptable. Although they are not pertinent directly to our subject
here, the data from the HNO$_3$-HCl leaching are of some general interest:
evidently for lake sediments — at least of the mineralogy in question — this
is a less good procedure than that with HNO$_3$ alone, although tests have
shown that on many soil samples only HNO$_3$-HCl leaching is effective. In
marine sediments, of course, the HCl equivalent is supplied by the interstitial
solution.

The IAEA's Monaco Laboratory [24] reported preliminary results of an interlaboratory comparison of plutonium analyses in a sample of marine sediment from Bombay Harbour. A variety of chemical attacks were used, ranging from HNO_3 leaching to aqua regia dissolution, to complete dissolution following fusion; it was clear that acid leaching was capable of giving good results for Pu in this matrix and that fusion was quite unnecessary.

For many purposes the easiest procedure to follow is to leach the wet sediment, measuring the weight loss on drying from a separate aliquot. Extra acid is, of course, required because of dilution by the interstitial water, but overall procedures go more smoothly and much faster.

5. RADIOANALYTICAL PROCEDURES

Methods are available for analysis in marine environmental samples of each of our five transuranics, except for californium. Some of these cannot be said to be fully developed, and several have been tested only on rather small samples of highly contaminated materials. Discussion of the methodological situation alphabetically: americium, californium, curium, neptunium and plutonium has the benefit that one concludes with the transuranic element that has been most studied, and about which there is the most to say.

5.1. Americium

In the present report Bojanowski et al. [22] describe a method for ^{241}Am that is believed to be the only one now being used routinely for measurement of this nuclide at fall-out levels in marine samples. This procedure is especially laborious because it separates the ^{241}Am from plutonium nuclides and because it must separate ^{241}Am also from all the stable lanthanides in the sample; seriously degraded alpha spectra result if the lanthanides are not removed. This latter precaution is especially necessary with sediments, which may contribute lanthanides in significant but variable amounts.

An alternative procedure, in use at Lowestoft [25] for ^{239}Pu and ^{241}Am in biological material contaminated by waste effluents consists of an ashing followed by HCl leach, reduction-oxidation steps to ensure equilibration with the monitors added, and then coprecipitation of the actinides with calcium oxalate. Generally plutonium and americium nuclides are determined together, by alpha spectrometry, following electroplating of the actinides after separation from other cations in $0.4\underline{M}$ HNO_3 and $4\underline{M}$ $NaNO_3$ with trioctyl phosphine oxide (TOPO) in n-heptane. Since ^{241}Am alphas cannot be resolved spectrometrically from those of ^{238}Pu, the presence of significant amounts of the latter nuclide, or the need to measure what is there, requires separation of americium from plutonium; this is done by oxidation of plutonium to the 4+ state, keeping americium in 3+, and extracting plutonium alone into TOPO from stronger $(2.0\underline{M})$ HNO_3.

The US AEC Health and Safety Laboratory Manual of Methods [26] recommends analysis of americium by initial precipitation of lanthanum fluoride, removal of Pu by thenoyl trifluoro acetone extraction, and finally removal of La and actinides other than Am as fluorides, while americyl fluoride is held in solution with $\underline{N}$ H_4F. The method has been used routinely by trained chemists on urine, mammalian tissues, faeces and air filters, but neither on any marine samples nor on any environmental samples at world-wide fall-out levels. For yield monitor [243]Am is used and scintillation spectrometry appears to be the preferred counting procedure.

5.2. Californium

As noted above, no attempt has been made to date to develop a procedure for analysis of californium in marine samples. It may be confidently predicted, however, from the work of Surls and Choppin [27] that californium nuclides would accompany americium in the thiocyanate procedure of Bojanowski et al.[22] and with probably somewhat better chemical yield. Californium-252, the isotope now being used in technology that might lead to its release to the oceans, is characterized by alpha particles of energy identical to those of [242]Cm. Should there be suspicion that samples contain both these nuclides, special procedures would be required. Either modification of the thiocyanate elution to separate californium from curium — indicated by Surls and Choppin [27] as difficult but not impossible — or analysis of [242]Cm by miliking its daughter [238]Pu, could be considered.

5.3. Curium

The americium method described by Bojanowski et al.[22] has proved (Livingston, private communication) to give also good recoveries of curium nuclides added as tracers. In a considerable series of tests the maximum change observed in Am:Cm ratio during separation was less than 15%. On the basis of these observations, Livingston et al. (to be published) have used [243]Am to monitor recovery of both americium and curium nuclides in environmental samples of water, organisms or sediments. Detectable amounts of [244]Cm have been found in only a few samples (among them IAEA seaweed sample AG-I-1 [8]), each probably representing a special contamination event. It should be noted that, as is notoriously the case with [239]Pu and [240]Pu, alpha spectrometry is unable to distinguish between [243]Cm and [244]Cm.

The method used at Lowestoft for americium would probably not separate these nuclides from those of curium, and careful, high resolution alpha spectrometry would enable the use of the Lowestoft procedure to yield data for curium of the same quality it delivers for [241]Am.

The Health and Safety Laboratory (Harley, private communication) is reported to employ a method for curium that, like that quoted above from Livingston, separates curium and americium together from the other actinides and all stable elements, and then differentiates by alpha spectrometry. Harley has suggested the use of [242]Cm for yield monitor in 'research'-level analyses of curium, and of [243]Cm or [244]Cm in the case of 'monitoring'-level analyses. Recent estimates [28], however, indicate that by the year 2000 the wastes from nuclear fuel reprocessing may have activity ratios of [242]Cm equal to [244]Cm; therefore, careful examination of the Cm:Am ratio uniformity during radiochemistry should be made to ensure that [243]Am can be used to monitor both actinides, or failing that, [248]Cm should be made available for yield monitor. The same estimates cited above indicate that [243]Cm in waste effluents will be about two orders of magnitude less abundant (activity basis) than either [242]Cm or [244]Cm.

Because of the low concentrations of ^{244}Cm likely to be encountered in environmental samples, special precautions must be taken to ensure the samples are completely cleaned of the lighter, naturally occurring actinides. In dealing with sediment samples Livingston has found it necessary to employ at least one extra clean-up step to remove the last traces of interferences.

5.4. Neptunium

Bojanowski et al. [22] have reported some steps that will make their americium procedure also useful for neptunium. So far (Livingston, private communication) they have analysed no environmental samples with detectable neptunium, although added neptunium tracer recoveries have been acceptable, if not gratifying.

Neptunium-237 has been measured (Noshkin, private communication) in some samples contaminated by close-in fall-out from the US Pacific Test Site. The ratio ^{237}Np to ^{239}Pu found in these samples was so low as to discourage strong efforts to measure neptunium in world-wide fall-out: it would appear to require scaling up sample sizes by at least a factor of 100.

5.5. Plutonium

5.5.1. Present status

As the transuranic element of greatest importance in nuclear technology and most widely disseminated in the environment, plutonium has naturally been the most studied in environmental samples. A considerable amount of this analytical literature has recently been reviewed [29]; perhaps one should not be surprised that in this review there is no mention of the marine environment. In fact, however, marine samples are among the earliest environmental materials whose content of plutonium was published [30 - 32], even though in none of these cases were the details of the analytical methods made available. In 1970 Wong et al. [20] published details of a method then in use for analysis of plutonium in seawater, marine organisms and sediments; later this was updated [33] and in the present report further revisions are described [34]. Also in the present report are presented the methods used by Aarkrog [19] and Pillai [35].

Each of these procedures is in current use for analysis of fall-out levels of plutonium in large volumes of seawater (∼50 litres), of marine organisms (kilograms fresh weight), or of marine sediments (50 - 500 g wet weight). Although the various laboratories involved have undertaken informal inter-comparisons — with generally satisfactory results — no formal interlaboratory comparison at these concentration levels has yet been completed; one is currently in progress with IAEA sponsorship, involving analysis of surface ocean seawater collected in the northern Sargasso Sea.

Several IAEA-sponsored intercomparison exercises have been completed, using small-volume aliquots of relatively highly contaminated seawater or algal tissue, while others are in progress. The results of the completed seawater and alga intercomparisons [8] show a rather large variety of methods in use among the 17 or more laboratories reporting. It appears likely that the generally small scatter of the data for ^{238}Pu and ^{239}Pu reported in these exercises may be attributable to a higher standard of housekeeping and detector calibration among laboratories willing to attempt methods as

complex and sophisticated as those for plutonium. Certainly the data show
that, at <u>monitoring levels</u> of contamination, a large proportion of laboratories
that undertake plutonium analysis obtain acceptable results. These results
should not, however, lead to euphoria: several laboratories, including some
that were successful participants in the IAEA intercalibration exercises,
have still been unable to scale up their procedures to deal with 50-litre
water samples, or with sediments at 0.4 dpm ^{239}Pu per kilogram range.
Many more intercomparisons will be needed, and especially at 'research
ranges' of contamination, to set the analytical problems posed by environmental plutonium in their proper context.

A number of general principles, caveats and annotations apply to the
separations methods now used for plutonium, those published or included
in the present report and those still unpublished.

5.5.2. Yield monitors

It should be categorically stated that <u>no</u> method available for plutonium
in environmental samples offers, or has any hope of offering, chemical
yields sufficiently reproducible to allow the omission of yield monitors.
These must always be used. Fortunately two convenient isotopes of plutonium,
^{236}Pu and ^{242}Pu, are available — or about to be available — in high purity for
use as yield monitors; neither is a significant constituent of any plutonium
mixture that causes environmental concern. In this report the profitability
of using both isotopes added at different points in a separations scheme is
demonstrated in pinpointing the steps at which losses are taking place [34].
In routine work only one monitor is needed, and the choice should be based
first on whether ^{238}Pu is to be sought, as well as ^{239}Pu and ^{240}Pu, and second
on how good is the preliminary estimate of the plutonium concentration in the
sample: ^{236}Pu has an alpha energy larger than that of ^{238}Pu, and ^{238}Pu, in
turn, exceeds that of ^{239,240}Pu. Since the amount of yield monitor added
should approximate to (or exceed) the level of the most abundant plutonium
isotope sought, and this is usually ^{239}Pu, there is real danger that the
degraded tail of the ^{236}Pu alphas, compounded by in-growth ^{232}U and ^{228}Th,
will contaminate the ^{238}Pu peak and decrease the accuracy of that measurement. In the cases of very low level samples, of inadvertent use of too
much ^{236}Pu, or of alpha spectrometers of poor resolution this effect may
even extend into the ^{239,240}Pu peak. In each of these cases, then, it will be
safer to use ^{242}Pu as yield monitor since its alpha energy is lower than that
of any of the plutonium isotopes sought and there is consequently no danger
of its peak bleeding into any other.

Although some effort, for other reasons, is now being put into making
^{237}Pu available in good radiochemical purity, this would offer help to the
yield monitor problem only if enough ^{237}Pu were added to permit its measurement by photon counting: the alpha energy of ^{237}Pu lies between those of
^{238}Pu and ^{239,240}Pu and is consequently even more likely to affect the accuracy
of measurement of ^{239}Pu, the most commonly sought isotope. Every effort
should be made to increase the availability of ^{242}Pu of very high radiochemical purity and of standardized disintegration rate.

5.5.3. Separations chemistry

In virtually every case plutonium procedures involve an initial collection
of plutonium on a carrier precipitate, after which it is separated from most

other radioactivity, and from all stable elements, by a series of ion exchange
and/or solvent extractions for final determination by alpha spectrometry.
The major difference, in principle, among the procedures in common use
appears to lie in whether they are designed for sequential application, so
that a single sample yields data for many radionuclides, or whether they
are designed uniquely for measurement of plutonium. As discussed especially
by Wong [33] and by Livingston et al. [34] in the sequential methods thought
must be taken to optimize the yields of each of the nuclides sought, often
with some sacrifice of yield of plutonium.

As noted by Fukai and Murray [8], in all but one of the methods described
in their plutonium intercalibrations the carrier precipitate was iron hydroxide,
and in most of these ferric hydroxide. Although Wong [33] advocated use of
ferrous hydroxide as the precipitate, Livingston et al. [34] were not able to
confirm his results and have abandoned the reduced iron carrier.

Pillai's bismuth phosphate procedure [35] seems to be capable of higher
and more consistent chemical yields than any other; a strong argument
against its wider adoption, however, is the difficulty of making it part of
a sequential scheme that would yield data for other nuclides, whether transuranics, fission products or activation products. Further effort should be
put into exploration of this possibility.

Recently Livingston et al. [36] have found that in their procedure the
iron hydroxide step can be eliminated, allowing all the nuclides usually
carried by it to come down with the acid calcium/strontium oxalate precipitate. This proved to save manipulation time and to lead to higher and more
reproducible yields. At the time of writing this report, in an effort to
include ^{55}Fe measurement in their sequential scheme, they are testing the
use of magnesium hydroxide to carry iron and the transuranics, but the
data available, though encouraging, are yet too few to support recommending
this procedure.

There is support both from the data of Livingston et al. [34] and from
unpublished work at HASL (Harley, private communication) for the argument
that higher and more reproducible yields of plutonium, especially in the analysis
of large volume seawater samples, can be obtained by use of larger amounts
of iron carrier. In most cases there appears to be no good argument against
such an increase — to, say, 5 g per 55 litre sample — routinely, other than
aesthetics: many analysts find so massive a precipitate of ferric hydroxide
simply distasteful. Certainly the anion exchange separation of iron and
plutonium, as outlined for instance by Wong [33], can easily be scaled up
to deal with any amount of iron. Of course, any sequential method of analysis
in which ^{55}Fe is being sought as well as plutonium requires limitation of the
iron carrier used, but otherwise the use of much increased amounts of iron
will be generally found beneficial to plutonium yields and reporducibility.

As noted above and in Fukai and Murray [8], the final stages of separation of plutonium from other alpha radioactivities, both in the natural
series and among the transuranics, may be either by ion exchange or by
solvent extraction. Experience at WHOI is that the ion exchange procedures
have been easier to adapt for analysis of very small amounts of Pu (fall-out
levels) in the presence of normal amounts of uranium or thorium and their
daughters. For monitoring levels, however, the solvent extraction procedures may well be quicker and cheaper (Lowestoft, private communication).
It should be noted in consideration of the several procedures [8] that
disregard this caveat, that the <u>final</u> separation of plutonium — or of any other

radionuclide for alpha-spectrometric determination — should be <u>without</u> carrier, with the purpose of yielding a counting sample as nearly weightless as is possible. Otherwise sensible degradation of the alpha spectrum will result, with reduction in resolution and consequent bad effects on the signal to noise ratio.

5.5.4. Source preparation

Procedures in common use call either for electroplating as the final preparation of plutonium for counting or for evaporation. Certainly the first is preferable, in terms of uniformity both of distribution of the source on the backing and of thickness of the deposit. Usually the reasons for adopting evaporation rather than plating are either cost of equipment, operator-time demanded, or uncertainty about the chemical yields of the plating step. Of course, in procedures that use solvent extraction for the final clean-up of the plutonium evaporation is a natural way of getting rid of the solvent, and it is tempting to prepare the source that way.

In the pyrazolone solvent extraction method [37] the final step is an evaporation of the organic phase that contains the plutonium. If the previous steps in the procedure have been carried out properly, it is possible to obtain a sufficiently thin sample for α-spectrometry by evaporation. However, only 2 ml of a total of 5 ml organic phase could be evaporated on the 2-cm diameter counting planchet. It was therefore necessary to divide the sample on two planchets each containing 40% of the total yield of the analytical procedure. In the case of a low analytical recovery of plutonium a single planchet would therefore contain only a minor fraction of the Pu-activity in the original sample, and the counting would be prolonged and inaccurate.

The evaporation method could therefore be recommended only if one deals with relatively high activity levels, or if the electroplating method gives low yields (e.g. less than 40%).

Low or uncertain yields in plutonium electroplating are a recurrent worry. Livingston et al. [34] note that after their procedure was changed to electropolate from ammonium sulphate medium and for longer times at lower current density (following the studies of Talvitie [38]) they achieved higher recoveries and better reproducibility. Pillai has had excellent results on a variety of marine samples by electroplating from HCl solutions onto platinum planchets, following studies by Mitchell [39]. He has observed that by carefully following the procedure it is possible to get consistently high electroplating efficiency. To check the electroplating efficiency for plutonium, the residual solutions after plating (including washings) were examined for plutonium content in a few cases. The results obtained by Pillai are given in Table VI.

The presence of iron in the plutonium fraction will adversely affect the count rate and the alpha spectrum of the plated sample. It is found necessary to evaporate to dryness the plutonium fraction with concentrated hydrochloric acid at least twice before transferring to the plating cell if steel planchets are to be used. One source of the plutonium loss in electroplating and consequent low recovery is due to absorption of plutonium by gaskets used in the cell. It is necessary to check this loss. Teflon gaskets are found to give no problem in this respect. Addition of reagents to the cell should be such that the plating bath will have a concentration of 0.2 g of chloride per ml and the total volume is 5 ml.

TABLE VI. EFFICIENCY OF ELECTROPLATING OF PLUTONIUM
FROM VARIOUS SAMPLE SOLUTIONS

Sample	Plutonium on plated sample (counts/min)	Plutonium in residual solution (counts/min)	Recovery by electroplating (%)
[236]Pu standard	31.6	0.2	99.38
[239]Pu standard	27.03	1.63	94.32
Organisms - 7	104.9	2.1	98.0
IAEA algae (52)	61.8	1.4	97.79
IAEA algae (90)	277.4	12.3	95.73
IAEA fine fraction	150.0	1.07	99.29
IAEA coarse fraction	106.00	5.17	95.35

Bowen and his co-workers agree that careful attention to detail is an important part of obtaining high and consistent recoveries by any of the plating procedures they have used. Of several analysts using the same equipment, following the same procedure, one or two (the most meticulous) will obtain consistently higher yields.

The use of polished stainless steel discs rather than platinum as electroplating planchets has other advantages beside the obvious financial one. Elimination of the pressure to clean planchets for re-use appears to Bowen the most important. As discussed below, the WHOI group obtain much useful data by milking the grown-in [241]Am, a year or so after plutonium plating, as a measure of the [241]Pu content of the sample. Both at HASL (Krey, private communication) and at Lawrence Livermore Laboratory (Noshkin, private communication) it has proved of value to have available a library of plated plutonium samples for recounting, for submission for mass spectrometry, or just to use for counter intercalibration. Although redissolved plutonium from planchet cleaning could be reserved for the same purposes, it is unlikely to be as the storage of solutions is less secure and more demanding than that of planchets.

It should be noted that an alternative exists to the procedures of plating or evaporating before alpha spectrometry. Procedures were worked out many years ago (Bowen, private communication) for the sulphonation of the surface only of preformed polystyrene sheets. These sulphonated polystyrene surfaces proved effective in the ion-exchange uptake of a variety of radio-active cations available as almost weightless samples, and were used in place of planchets in several different counting modes. Since the sources so prepared were essentially monomolecular films and the radioactivity was very evenly distributed, they were especially desirable for the counting of low energy beta or of alpha radiation. Although no recent application of this technique has been known, it seems likely that on a routine basis it would compete favourably in terms of cost and reliability with plating, and be generally superior to evaporation.

For really precise work at low levels careful attention must be paid
to the interval between sample preparation and introduction to the alpha
spectrometer. Bowen and coworkers (Mann, private communication) allow
at least 14 days after plating of samples to be counted for ^{238}Pu or ^{241}Am,
to allow any ^{228}Th that may have accompanied the nuclide sought to come
into equilibrium with ^{224}Ra, ^{220}Rn and ^{216}Po; the very high energy alphas
of these (5.68, 6.29 and 6.78 MeV, respectively) are well distinguished
from those of any of the artificial nuclides sought, and can be used to
estimate for subtraction the ^{228}Th contribution to the peak, about 5.50, that
is attributed to ^{238}Pu or to ^{241}Am. It is comparably important, however,
not to wait too long between the final separation of plutonium and its alpha
spectrometry, since the ^{241}Am that grows in from decay of ^{241}Pu increasingly
contaminates the ^{238}Pu peak. In some waste disposal situations, especially,
the ^{241}Pu concentration is quite high [40] and very careful time control may
be needed for accurate work.

A final note applies to the measurement of ^{241}Pu, the beta-emitting
isotope of plutonium. Relatively simple energy discrimination applied to
liquid scintillation counting [40] allows the determination of these soft betas
(0.021 MeV maximum) in the presence of large amounts of any of the plutonium
alpha emitters. The relatively high backgrounds that are seemingly unavoidable in liquid scintillation prevent this technique from competing, in
absolute sensitivity, with alpha spectrometry. In a variety of monitoring
situations, however, the limiting sensitivities available (about 3 picocuries)
for ^{241}Pu by liquid scintillation may be quite sufficient, and then the technique
should be seriously considered. In such cases a final solvent extraction step
is indicated, using any materials that are compatible with the standard
scintillation cocktails.

5.5.5. Detection methods

It has been noted just above that in special cases liquid scintillation
may be useful for measurement of ^{241}Pu; in mixtures so simple that its
relatively poor energy discrimination is adequate and with enough plutonium
so that background levels are no problem this has been useful for alpha
measurement, too [41].

For samples of high enough levels of activity the use of thin (2 - 3 mm)
crystals of thallium-activated sodium iodide permits determining ^{239}Pu by
counting its 17 keV X-ray — or ^{241}Am by its 60 keV gamma ray — (Lowestoft,
private communication). Sensitivities are low enough so that only in the
cases of sediments close to waste-stream outfalls, or of serious accidents [10]
is this approach useful.

Before proceeding to discuss the various electronic spectrometric
methods in use for measurement of plutonium alphas, some other techniques
should be noted: The use of cellulose nitrate films, etched after exposure,
for alpha particle counting; and of fission-track counting, using either mica
or lexan sheets for registration. Sakanoue et al. [42] have shown that both
of these methods are useful, especially for the 'rapid' measurement of
plutonium in environmental samples. The latter method, although it requires
access to a neutron source, has the advantage, as will be discussed later on,
of being potentially selective for the plutonium isotopes of high fission cross-
section (^{239}Pu and ^{241}Pu). Both methods have the advantages of using only

inexpensive materials and equipment and, by extending the times of exposure, of very considerable sensitivity. Neither method, unfortunately, is at all selective for plutonium and both, especially the fission track procedure, place great demands on the radiochemist to be certain that his separation scheme has eliminated all traces of other fissionable or alpha-emitting nuclides.

An alternative to the cellulose nitrate film procedure is the use of nuclear track emulsions for the photographic registration of plutonium alphas. Photographic detection has several advantages, probably the most salient being the ease with which alpha tracks can be traced to their source. This permits the identification, and even the location, of point sources of high rates of alpha emission and has been much used (Harley, private communication) in the study of plutonium distribution on and among aerosol particles. It is possible to achieve a respectable degree of energy discrimination by measuring the length of alpha tracks registered on nuclear emulsions; it is furthermore possible, in some special cases, to use the incidence of two, or more, track 'stars' as a measure of the relative abundance of the short-lived plutonium isotopes, or of the frequency of aggregations of large numbers of plutonium nuclei. With photographic exposures of respectable length it should be possible to ascertain from the presence of multiple track stars how seriously a 'plutonium' sample may still be contaminated with natural alpha emitters that have several short-lived daughters. It appears that use of photographic detection procedures should permit a good deal of very profitable and fundamental research to be done by people who cannot afford to equip themselves with alpha spectrometers.

By far the mass of plutonium and transuranic research has been carried out using electronic detection devices. For total alpha measurement, and in some areas this can be equated, after radiochemistry, with total plutonium measurement, gas-flow proportional counters and scintillometry using zinc sulphide phosphors coupled to photomultipliers have been used (Lowestoft, private communication); in configuration like the latter thin films of plastic scintillators are also useful (Harley, private communication). To achieve really low backgrounds, and the kind of energy resolution required for use of Pu isotopic yield monitors, true alpha spectrometry is required.

For alpha spectrometry detection is either by semi-conductor solid-state diodes (either gold or, more recently, aluminium-treated silicon) or Frisch grid chambers. The Frisch grid chamber and the semi-conductor detector are both excellent for alpha spectrometry. The advantages of the chamber are its high efficiency, approaching 50%, and its ability to handle large samples. One disadvantage is that the sample must be conducting. The diode has slightly better resolution in most cases but the usual efficiency is in the range from 15 - 25% and they are more difficult to clean if they become contaminated. Both detectors are relatively complex in terms of inserting the sample, since most diodes are operated under vacuum or with a low Z gas, and the gridded chambers are operated as flow counters and it its necessary to flush the chamber after putting in the sample. In both cases the signal from the detector can be handled adequately by standard electronics, and a 100-channel analyser is sufficient for data accumulation if it can be restricted to the range 4.5 to 6.5 MeV energy equivalent. Probably the biggest advantage that the semi-conductor detectors have is their wide commerical availability, and being very much in fashion.

Counter contamination has proved a real and a serious problem and the possibility of decontamination of the inner surfaces of a Frisch grid chamber is an important advantage; although the newer aluminized-silicon semi-conductors can be wiped clean, this does not offer real decontamination, as discussed below. Surface contamination may be removed from stainless steel surfaces by electropolishing. This process actually removes a thin surface layer by making the object to be cleaned the anode in an electrolysis. It has been used successfully on many counter components and interior surfaces of grid chambers to reduce background.

The stainless steel object to be cleaned is made the anode in a bath of concentrated phosphoric acid. The cathode may be metallic but a carbon rod works very well. The current should be from 1/2 to 1 ampere per square centimetre of surface and electrolysis should continue for 1 or 2 minutes. The part can then be washed thoroughly in running water and dried. This latter step can be speeded by rinsing in alcohol.

The current requirements are very considerable and are not generally available in laboratories. A small d.c. welding machine is a suitable current source.

5.5.6. Sources of detector contamination

Relatively recently, when it became known that most manufacturers of semi-conductor alpha detectors used uncovered ^{241}Am sources for energy calibration, this practice was argued to be the source of the very commonly observed contamination of such detectors by a nuclide emitting alphas at about 5.5 MeV energy. Such contamination interferes especially with the measurement of ^{238}Pu and many laboratories, to avoid the supposed con-tamination problem, arranged to purchase at their own risk uncalibrated detectors. This is referred to by Fukai and Murray [8]. In fact, however, uncalibrated detectors soon, and widely, proved also to be contaminated, and it was apparent that the answer lay in the report by Sill and Olson [43] of the dangers and mechanism of detector contamination by recoil of radio-active daughters in the natural decay chains. This is an unavoidable con-comitant of alpha spectrometric analysis of mixtures of uranium or thorium isotopes and their daughters, but in the cases of detectors reserved for transuranic analyses probably arose most commonly from use, in the early stages of setting up and calibrating new systems, of too intense sources of ^{236}Pu in at least partial equilibrium with its ^{232}U daughter, whose decay in turn recoils ^{228}Th onto the detector surface. In Bowen's laboratory (Livingston, private communication) long background counts of detectors showed essentially all the alpha peaks to be expected from this source. Both Sill and Olson [43] and an earlier, generally ignored, report by Chetham-Strode et al. [44] describe procedures for eliminating or greatly reducing recoil contamination of detectors by operating them under gas pressures that exceed the mass absorption coefficients of the recoil atoms but that cause little degradation of the alpha spectra. Sill and Olson found that there was a significant drift of the resting recoil atoms to the detector because of the charge on the latter; this they could eliminate by maintaining the source plate at a slight negative potential, as little as 2 volts. These procedures have been tried in Bowen's laboratory (Livingston, private communication) and have proved effective, but by no means convenient.

Preliminary experience in Bowen's laboratory seems to him to indicate that the wipeable, aluminized silicon semi-conductors could be cleaned of contamination by the drifting recoil atoms, but not of contamination by recoil atoms that strike the detector with most of their recoil velocity.

For a laboratory engaged principally in transuranic research, Bowen's coworkers have found (Mann and Livingston, unpublished) that simple good housekeeping is sufficient, following only a few rules:

(a) Keep all standard radioactivity sources to minimum activity levels — 2 - 5 dpm maximum — and use generally only freshly separated ^{236}Pu sources.

(b) Whenever possible use ^{242}Pu as yield monitor, since it decays to ^{238}U, which has so long a half-life that its further decay on the detector is insignificant.

(c) Examine all sample spectra after only a few minutes of counting and again after an hour or so, to ensure that no really high level sources are left in the counting chambers for long periods and that any sources showing significant contents of the natural alpha-decay chains are removed at once for further decontamination.

As a result of following these precautions rigorously their 10-detector system has shown no evidence of background build-up over the past 18 months, whereas previously they had to reserve detectors of 12 - 18 months service only for high-level environmental samples because their backgrounds had become so high.

A certain amount of confusion appears to exist, especially among investigators new to semi-conductor spectrometry, about the resolution to be expected in using silicon detectors for alpha spectrometry. The usual manufacturer's procedure in determining the detector resolution specified is to calibrate with a source-to-detector separation equal to the detector diameter, and with a source that is small in diameter compared to the detector. This comes close to giving maximum resolution, being enough separation so that most alphas strike the detector at angles greater than 60° and consequently have a high probability of dissipating all their energy in the detector's sensitive volume, and yet not a great enough separation to allow the detector to see a significant number of alphas that have been scattered off the sides of the counting chamber with consequent loss of energy. Unfortunately, at separations such as this the geometric efficiency of detectors is low; most environmental samples are counted, in practice, very close to the detector (1 - 2 mm) to optimize counting geometry, but with a loss in resolution. Careful measurement of this effect in Bowen's laboratory, using a source of 13 mm diameter, (Mann, unpublished) showed that an Ortec detector (300 mm^2 by 100 μm) gave 54.2 keV full-width half-maximum resolution, and 28.4% efficiency at 1.1 mm source-to-detector separation (standard for this detector), but at 11.5 mm separation gave 35.6 keV resolution and only 8.7% efficiency; at large separations resolution was somewhat less good in the counting chamber used. The manufacturer's specified resolution with this detector was 16.0 keV. Misunderstanding of this situation seems to be responsible for a good deal of dissatisfaction, especially with newly set-up systems, and also for a certain number of failures to meet special circumstances demanding the highest resolution by increasing the source-to-detector separation, and accepting that this requires a much increased counting time.

5.6. Plutonium isotope ratios

As noted above, the most commonly used detection method for plutonium, alpha spectrometry, is incapable of distinguishing between ^{239}Pu and ^{240}Pu because their alpha energies are too close to be resolvable; it is also, by definition, incapable of detecting ^{241}Pu, a pure beta emitter. Because of the single injection of ^{238}Pu from the SNAP-9A burn-up, a good deal of useful information is obtainable from the ^{238}Pu to ^{239,240}Pu ratios that can be obtained by alpha spectrometry. But a good deal more useful information lies in the ratios of ^{239}Pu to ^{240}Pu, and of ^{239}Pu to ^{241}Pu, that we mostly miss.

Recently some data of this sort have been obtained with mass separators especially cleaned up to deal with the very small amounts of total plutonium available in environmental samples [45 - 47]. It is evident that specific sources of plutonium may be characterized by unusual or unique plutonium isotope ratios, which may then be useful to identify the fraction of contamination from such a source, referred to the contamination from 'global fallout'. It is also evident that whenever the source ratios of ^{238}Pu (half-life 87.8 years) or of ^{241}Pu (half-life 15 years) can be known, their changes with time should provide useful dating information; comparably, of course, demonstration of uniformity of plutonium isotope ratios (as seems to be indicated for some soil profiles, Volchok, private communication) would provide confirmation of rapid vertical mixing processes. Finally, it should be noted that since the decay of ^{241}Pu results in production of ^{241}Am, interpretation of differences in the ratio of this nuclide to ^{239,240}Pu is difficult without knowledge of the ^{241}Pu content of the plutonium fraction.

Unfortunately, mass spectrometers capable of analysis of plutonium isotopes are rare, very expensive, and demand rather larger amounts of plutonium than the environment easily yields. Moreover, such instruments are usually engaged in the analysis of isotope ratios of very large amounts of plutonium (by environmental standards), often of rather bizarre isotope ratio, so that the dangers of cross-contamination are very real, and their avoidance demands much time and skill.

For selective measurement of ^{241}Pu two alternative methods are available, both of which have been used:

(a) The beta decay of ^{241}Pu can be measured in liquid scintillation, with pulse height analysis to discriminate against the alpha decays [40]. This procedure has yielded useful data on soils and waste streams, but beta detector backgrounds are always large compared to those of alpha spectrometers so that extension of the method to low-level environmental specimens will require collection and processing of more sample than one likes to deal with.

(b) The americium method described by Bojanowski et al.[22] offers an approach to measurement of ^{241}Pu: the plutonium plated for alpha spectrometry needs simply to be reserved for a period sufficient to allow the growth of a measurable amount of ^{241}Am. The alphas from the ^{241}Am may be counted directly on the plutonium plate, as in the HASL procedure (Harley, private communication), or the plutonium may be dissolved, and the americium, separated and plated [48], counted separately. Since the alpha energy of ^{241}Am (5.49, 5.54 MeV) is not resolved from that of ^{238}Pu (5.50, 5.46 MeV), counting of the separated americium permits measurement of smaller amounts with more precision than does its counting simply as an increment to the ^{238}Pu peak. This is an even more serious problem when excess ^{236}Pu used as yield monitor has contaminated the ^{238}Pu peak.

Using the ^{241}Am milking procedure, after decay periods of 1 - 3 years Livingston et al. [48] have obtained ^{241}Pu to ^{239,240}Pu isotope ratios on a number of marine samples contaminated variously by world-wide fall-out, by accidental plutonium release, or by disposal of reprocessing waste. Unfortunately, very long waiting times and very large sample sizes appear to be required for application of this method to most marine biota.

For measurement of ^{240}Pu to ^{239}Pu ratios, it appears that use might be made of the great difference in the fission cross-sections of the two isotopes (742.4 barns for ^{239}Pu versus <0.1 barn for ^{240}Pu). By plating and counting the combined alphas in the usual way, and then applying to the plated sample the fission track procedure described by Sakanoue et al. [42] one would obtain a measure of $^{239+240}$Pu first and then a measure of ^{239}Pu alone. In such a procedure it would be important to use, rather than ^{236}Pu, ^{242}Pu as yield monitor because of its much lower fission cross-section (<0.2 barns versus 170 barns for ^{236}Pu). The indication from Noshkin and Gatrousis [47] that environmental ratios ^{240}Pu to ^{239}Pu may range from 0.1 to 0.27, and possibly as high as 0.36 for pure debris of one 1958 test [49], suggests that careful application of a procedure such as outlined should produce useful information. It seems that this has not yet been attempted. It is curious that Sakanoue et al. [42] seem to have ignored the fact that ^{240}Pu was measured by their alpha-track method but not by their fission-track method. Their very high ratios of ^{238}Pu to ^{239}Pu suggest that something else may also have been wrong in their experiments; the spectra they figured show this was almost certainly inadequate decontamination from ^{228}Th and its daughters.

In conclusion, it appears that radiochemical procedures are available for measurement of each of the four 'environmental' plutonium isotopes, and wider examination of their ratios in environmental samples should be urged. This effort would be benefitted by availability of intercomparison samples with plutonium at environmental levels and with known, and different, isotope ratios. Our understanding of the sources and movements of environmental plutonium would certainly be greatly advanced should it be possible for the IAEA or some appropriate government to make available mass-spectrometer capability for as many as 100 environmental plutonium isotope ratio measurements each year for several years.

6.   PRESENTATION OF DATA

In summarizing and tabulating data on the results of analytical measurements, it should be remembered that the data may be useful to others in different fields for other purposes. To enhance this potential usefulness, the presentation should include all of the primary information (e.g. wet, dry and ash weights, yields, sample activities) as well as the descriptive information (e.g. sampling method, analytical scheme, counting procedure). Such material is probably a natural part of any data presentation. Other equally important, but frequently omitted information should also be included,

such as calibration methods, procedural problems and statements as to the
reliability of the reported results.

In describing the calibration of the measurements, both the source of
the standards and the supplier's specifications should be mentioned. Other
pertinent information would include any treatment of the standard prior to
analysis (such as dilution, mounting or plating) and the specific details of
the measurement methods employed.

There are a number of frequently encountered procedural problems
that cause systematic biases in the data. One example is the existence of
a significant blank. This is generally a contamination of some sort, perhaps
in the sampling system (filter paper, ion exchange resin), in the analytical
reagents, in the laboratory or maybe even emanating from the personnel.
The source or sources should be explored and discussed in the data presentation, and an estimate of the variability given. Whether or not a blank was
subtracted from the presented data should also be mentioned.

The reliability of the results should be thoroughly discussed. For this
a complete description of the methods used to evaluate and control the quality
of the analyses is desirable, followed by a summary of the quality control
results. For estimating the accuracy, standards should be analysed blind
(disguised as real samples). For precision blind duplicates are required;
repeat analyses of the same standard also serve in evaluating precision.
Finally the standard deviations tabulated with the presented data must be
clearly defined as either the Poisson counting error or the Gaussian error
representing a statistical evaluation of a stated number of replicate analyses.

Although discussed in another paper in the present report [50], it should
be emphasized here also that only disaster can be the final result of neglect
of routine quality control analyses. This is especially a danger in the low-
level analysis of transuranic nuclides because each analysis is difficult,
time-consuming and expensive, with the result that management is likely
to wish to maximize output of unknowns and minimize quality control. The
arguments often stated: "Oh, we did well in last year's intercomparison",
or "We just are too busy turning out data to engage in quality control",
have no validity. Innumerable sources of error lie in wait constantly for the
radiochemist engaged in these programmes, and they can be seen — and
eliminated — only be constant attention to the data from many kinds of quality
control. At the same time, data that are unaccompanied by quality control
information are really not data at all but rather sorts of floating 'guesstimates'
that become interpretable only in the light of intercomparisons of proven
relevance. A salient problem in this context is the reluctance of most
journals to publish adequate detail of the quality control experience that is
needed to support the data being reported. Obviously stronger pressure is
needed from the members of the scientific community to improve this
situation, but as publishing costs increase alternative ways to provide this
information may have to be found.

In conclusion, it should be also emphasized that although the term
'standards' is employed above — and indeed standards are urgently needed
by most environmental programmes — in fact standards in the strict sense
are not available. The samples distributed by the IAEA for interlaboratory
comparison analysis [6 - 9] have become quite well known, but really only
in the sense of a statistical concensus. This is even more true for the
quality control samples used in the programmes described by Harley et al.

[50]. Great benefit would accrue to the whole field of environmental radio-activity if a group of government or international laboratories were to produce a series of true standards, of several matrices at several levels of activity, and ideally accompanied by certified blanks, of the same matrices.

III. RADIORUTHENIUM

1. INTRODUCTION

The radionuclides of ruthenium, ruthenium-103 and ruthenium-106, have been known to be present in fresh radioactive fall-out as a result of nuclear tests and, as a result of its longer half-life, ^{106}Ru constitutes one of the major radionuclides in fall-out aged from a few to several years. On the other hand, since ruthenium has a complicated chemistry exhibiting several valency states between 0 and 8 and forming strong complexes, it tends to escape various steps of waste effluent treatments and find its way into the aquatic environment. In fact, ^{106}Ru has been one of the major radionuclides of waste effluents from several nuclear installations, especially from nuclear fuel reprocessing plants [51 - 55]. Thus, the measurement of ^{106}Ru in the aquatic environment is important in monitoring operations of many nuclear installations.

Recent results of intercalibration exercises organized by the IAEA of the analytical methods for radionuclide measurements on marine environmental samples such as seawater [6], seaweed [9] and marine sediment [7] have demonstrated, however, that the comparability of the results of ^{106}Ru measurements performed by different laboratories on these samples has not been always satisfactory. The scatter of the results reported for ^{106}Ru in these intercalibration exercises can be indicated by the ratio of maximum value reported to minimum value (scatter index), which varied from 30 (seaweed) [9] to 10 000 (marine sediment) [7]. Since the majority of the participating laboratories in the seaweed and sediment intercalibrations employed non-destructive γ-spectrometry with Ge(Li) detectors, the large scatter of the reported results indicates that there are still problems in calibration of the γ-spectrometer and/or radionuclide standard used. On the other hand, the large scatter indices for the results of the seawater intercalibration, that is 50 for lower level seawater and 200 for higher level seawater [6], suggest that several radiochemical procedures adopted by different laboratories may be the main cause of the scatter in this case. Since radiochemical methods of analysis for ^{106}Ru in marine environmental samples have been comprehensively reviewed by one of the supporting papers of this report [56], general discrepancies of the experimental results obtained between different workers are pointed out and discussed here. It appears that these discrepancies were mainly caused by deducing conclusions from experiments for which different chemical forms of ruthenium were involved. For this reason it seems essential that analytical workers have at least some understanding of the chemical behaviour of ruthenium in the marine environment in order to choose suitable analytical procedures for their work. The present status of the knowledge of the chemical behaviour of ruthenium in seawater is also outlined below.

2.   CHEMICAL BEHAVIOUR IN THE MARINE ENVIRONMENT

2.1. Nitrosyl ruthenium complexes in effluents

After the treatment of irradiated fuels with nitric acid, ruthenium is present in the form of complexes of nitrosyl ruthenium, RuNO, which is very stable and has an electronic charge of 3+, and in which ruthenium has a co-ordination number of 6. Two groups of complexes are found: nitrato complexes and nitro complexes.

The nitrato complexes can be expressed by a general formula

$$[RuNO(NO_3)_x (OH)_y (H_2O)_z]$$

whose $x + y + z = 5$. In nitric acid solution above $0.1\underline{N}$ it is improbable that the OH group occurs, so that the general formula can be rewritten as follows:

$$[RuNO(NO_3)_x (H_2O)_{5-x}]^{3-x}$$

It seems well established that nitrato complexes are under the following equilibria in nitric acid solution [57, 58],

$$[RuNO(NO_3)_5]^{2-} \rightleftarrows [RuNO(NO_3)_4(H_2O)]^- \rightleftarrows [RuNO(NO_3)_3(H_2O)_2]^0 \rightleftarrows$$
Penta-nitrato complex   tetra-nitrato complex   tri-nitrato complex

$$[RuNO(NO_3)_2(H_2O)_3]^+ \rightleftarrows [RuNO(NO_3)(H_2)_4]^{2+} \rightleftarrows [RuNO(H_2O)_5]^{3+}$$
di-nitrato complex   mono-nitrato complex   non-nitrato complex

The relative proportions of these nitrato complexes vary depending on the concentration of nitric acid. With decreasing acidity these forms become unstable and tend to hydrolyse.

The nitro complexes include at least one NO_2 group and can be written [58] as

$$[RuNO(NO_2)_x (NO_3)_y (OH)_{3-x-y} (H_2O)_2]$$

where $x \geq 1$. They are much more stable than the nitrato complexes and are difficult to destroy once they are formed. There is evidence [59] that the di-nitro complex is under equilibria among the following three forms:

$$[RuNO(NO_2)_2 (OH)_2 H_2O]^- \rightleftarrows [RuNO(NO_2)_2 OH(H_2O)_2]^0 \rightleftarrows$$
anionic complex   neutral complex

$$[RuNO(NO_2)OH(H_2O)_2]^+$$
cationic complex

2.2. Nitrosyl ruthenium complexes in seawater

Experimental studies on the behaviour of nitrosyl ruthenium in seawater have been carried out in several laboratories. Since these experiments were, however, done under different conditions, it is difficult to compare the results obtained by one laboratory with the others. The

acidity of the nitrosyl ruthenium before it was introduced into seawater
varied from $0.1\underline{N}$ to $10\underline{N}$, depending on the laboratories, and the time
during which the nitrosyl ruthenium was stored at a specific pH was not
generally indicated. These factors influence substantially the behaviour
of nitrosyl ruthenium in the seawater medium. For example, at an acidity
of $0.1\underline{N}$ polymerization can become very important [57, 60], and nitrato
complexes may possibly be hydrolysed and present in forms such as
$[RuNO\text{-}O\text{-}RuNO]^{n+}$ or $[RuNO(OH)_2(H_2O)_2\text{-}O\text{-}RuNO(OH)_2(H_2O)_2]$ [61].
These polymerized forms are considered very stable and it is necessary
to boil in $8\underline{N}$ HNO_3 to destroy them [62]. Nitrato complexes in $0.1\text{-}1\underline{N}$
acid solution are probably in cationic forms, mono- and di-nitrato com-
plexes [57, 63]. At an acidity of $8\underline{N}$ no polymerization will occur and
ruthenium is essentially in anionic forms as penta- tetra- and tri-nitrato
complexes [57, 63].

To give an example of the partitioning of various forms of nitrosyl
nitrato complexes in the process of introduction of ruthenium into seawater, a series of experiments carried out by Guéguéniat (private communication) are described below. In these experiments the partitioning
of complexes was measured by passing the solution through an ion exchange
column [64]. When the solution of nitrosyl ruthenium nitrato complexes
(0.1 mCi in $6\underline{N}$ HNO_3) was added to a small volume of seawater (50 ml),
the pH of the seawater was lowered from 8 to 1. After partial neutralization with NaOH solution, the pH of the seawater was increased to 6 (the
neutralization operation took around 10 min). Immediately after neutralization the acidic seawater was diluted with 2 litres of seawater, the pH of
which eventually returned to the original value of 8. The partitioning of
nitrato complexes was measured at three steps described above. The
results of these measurements are presented in Table VII. As can be seen
from the table, the polymerized fraction of nitrosyl ruthenium increases
dramatically with increasing pH. The polymerized fraction is considered
to serve as primary nuclei for colloid and particulate formations. The
products thus formed may have either positive or negative electric charge,
which influences substantially the behaviour of colloids or particulates.
Positively charged particles tend to be absorbed by certain seaweeds or
walls of containers within a relatively short time, while after several weeks
negatively charged particles tend to remain in the medium (Guéguéniat,
private communication). In this way polymerization of ruthenium nitrato

TABLE VII. PARTITIONING OF NITROSYL RUTHENIUM NITRATO
COMPLEXES IN SEAWATER AT DIFFERENT pH VALUES

	at pH 1	at pH 6	at pH 8
Penta-, tetra-, tri-nitrato complexes	10-11%	7%	4%
Di-nitrato complexes	54-55%	12%	11%
Mono-nitrato complexes	30-31%	24%	17-18%
Polymerized	4%	55%	66%
Undetermined	3%	2%	1-2%

complexes and successive colloid and particulate formation affect radiochemical recovery of ruthenium depending on the history of seawater
samples. Thus, the analytical procedures, which are capable of analysing
only certain forms of ruthenium, may miss considerable fractions of
ruthenium to be measured.

The chemical behaviour of nitro complexes of nitrosyl ruthenium has
been studied much less than that of nitrato complexes, despite the fact that
nitro complexes are considered to be a major portion of the ruthenium
released from nuclear plants. They are particularly stable and not hydrolysed easily, while they are partially precipitated with certain stable
elements occurring in waste effluents when the effluents are brought into
contact with seawater.

In conclusion, further studies on the equilibria between different soluble
and colloidal forms of ruthenium in the seawater medium are urged. These
equilibrium reactions should be considered as one of the major governing
factors of contamination kinetics of organisms with radioruthenium in the
marine environment.

3. DETERMINATION IN SEAWATER SAMPLES

The normal practice used in monitoring radioactive wastes in the
marine environment is to measure radionuclides in the highest step of
critical pathways, such as marine organisms (directly consumed by man)
and beach sand (contributing to the external dose received by man).
Although seawater does not usually represent the highest step of critical
pathways, it has been estimated that the required lower limit of the determination of ^{106}Ru in seawater for monitoring purposes should be
0.2 pCi/litre [5]. This value was arrived at by considering concentration
processes of ^{106}Ru by various marine organisms. It is close to some
results obtained for seawater samples collected in areas not influenced
by radioactive waste effluents. Since the limit of detection for in situ
γ-spectrometry of ^{106}Ru is known to be from 35 pCi/litre [65] to
70 pCi/litre [66], depending on the γ-spectrometer set-up, in situ methods
for monitoring are limited to special circumstances, such as in the
proximity of waste effluent outfalls. Even a laboratory set-up of a conventional γ-spectrometer is not sufficiently sensitive for direct measurements of ^{106}Ru in seawater samples since it requires 50-100 pCi of ^{106}Ru
per sample for its quantitative determination. Although the sensitivity of
a γ-spectrometer can be improved with more sophisticated instrumentation,
this expense is usually not justified for monitoring operations. Consequently,
some preconcentration steps for ^{106}Ru in bulk seawater have to be used for
its routine measurements to achieve the stated lower limit of determination required for monitoring.

The various preconcentration and radiochemical separation procedures for radioruthenium can be divided into two categories, that is,
procedures using added stable ruthenium carrier and those without the
addition of the carrier. The former procedures are suitable for higher
activity-level samples of a size from a few to several litres, while the
latter procedures are normally used for low-level samples of more than
50 litres in volume. Since the sensitivity of β-counting is about one order
of magnitude higher than that of γ-spectrometry (minimum quantity of

[106]Ru required for β-counting being approximately 5-10 pCi as against
50-100 pCi needed for γ-spectrometry), β-counting is more commonly
used after radiochemical separations.

3.1. Procedures with added carrier

The definite advantage of carrier procedures is that they make it
possible to determine directly the chemical yield of radioruthenium by
measuring added stable ruthenium, recovered after radiochemical separation procedures. As discussed in the preceding section, however, the
complicated behaviour of ruthenium in the seawater medium does not make
the precise yield determination an easy task. Radioruthenium introduced
by waste disposal operations into seawater may be present as a mixture
of nitrato and nitro complexes of nitrosyl ruthenium in various valency
states and their polymerized forms, the partitioning of which changes
depending on the time of contact of the effluents with seawater. On the
other hand, stable ruthenium in carrier solutions to be added to sample
waters is usually in higher valency states of ruthenium with added oxidizing
agents to maintain ruthenium in solution. For this reason it is essential
to achieve complete isotopic exchange, by some means, between the radioruthenium originally present in seawater samples and the added ruthenium
carrier, in order to determine radiochemical yield of radioruthenium by
the stable carrier measurement. Although the equilibria and kinetics of
various ruthenium reactions in the seawater medium have not yet been
fully understood, it is empirically supported that vigorous oxidation brings
practically all the ruthenium present in seawater into higher valency states,
possibly perruthenate, and achieves the necessary isotopic exchange between
radioruthenium and stable carrier. It has not been demonstrated, however,
what fraction of ruthenium is in forms that resist oxidation, e.g. well-
polymerized complexes or large-molecule organic complexes. Further
experimental studies are needed in this respect.

With these remarks in mind, it can be said that oxidation with potassium
persulphate and periodate with heating, adopted by Loveridge [67], WHO [68]
and Iwashima [69] among others, seems to bring at least substantial portion of ruthenium into the perruthenate state. Although the conditions of
oxidation described in these references are slightly different, the key point
is to achieve the oxidation with sufficient heating time, without losing
volatile perruthenate into the atmosphere. Longer heating times than those
described in the literature might sometimes be necessary to ensure oxidation.

After the oxidation step, ruthenium is separated from other radionuclides by either solvent extraction or distillation. Although the distillation is a sure method of separation, the solvent extraction procedures are
more commonly used because of their simplicity. Finally, ruthenium is
again reduced to dioxide or the metallic state to be mounted for counting
and yield determination.

For the chemical yield determination two different methods, spectrophotometry of perruthenate and gravimetry of ruthenium dioxide or metal,
are usually used. Although the papers already referred to adopted spectrophotometry, it seems that gravimetry of ruthenium metal has somewhat
better reproducibility and accuracy.

It should be emphasized that at present the oxidation-reduction cycle
described above seems to be the most reliable and practical method for

monitoring seawater, provided the isotopic exchange between radioruthenium and its carrier is well performed. The method is, unfortunately,
not practical for treating large seawater samples.

3.2. Procedures without addition of carrier

Although the procedures in this category are more easily applicable
to large volume seawater samples than those in the previous category,
the obvious drawback is that it is not possible to determine the chemical
yield of radioruthenium by measuring the recovery of the stable carrier
added. Therefore, the preconcentration procedures should collect 100%
or nearly 100% of radioruthenium from seawater samples in order to
obtain reasonably accurate results. Although some methods proposed to
date are successful for recovering approximately 100% of the radioruthenium from ruthenium tracer solutions prepared in the laboratory,
they are incapable of doing so from seawater collected in situ because of
the different chemical forms of radioruthenium involved. For example,
Guéguéniat et al. [70] tested the coprecipitation of radioruthenium with
manganese dioxide directly precipitated in the sample medium and arrived
at the following results for the various samples tested:

Ru added in chloride form,	100% Ru recovery
Ru added as nitrato nitrosyl complexes,	100% Ru recovery
Ru in the effluent from the La Hague Plant (1966-69)	100% Ru recovery
Ru in effluent from the La Hague Plant (1972-73)	45-65% Ru recovery
Ru in the effluent from the Marcoule Plant	50-55% Ru recovery
Ru in the intercalibration sample SW-I-1	100% Ru recovery

As shown by these results, the recovery of radioruthenium by coprecipitation with manganese dioxide varied, depending on the source of
ruthenium. In these experiments it was noticed that the recovery increased
for ruthenium in the effluent from the La Hague Plant (1972-73) from approximately 55 to 90-95% when the coprecipitation was made with heating at
about 95°C. The recovery of radioruthenium was also raised by prolonged
contact time; after 1 day it was 55%, after 1 week 84% and after 2 weeks
93%. These observations suggest that the equilibria between various forms
of radioruthenium were displaced in such a way as to form adsorbable
species during the prolonged contact time, after certain forms of radioruthenium had been fixed onto manganese dioxide precipitate. Heating the
system seems to accelerate this reaction.
    On the basis of the results of the above-mentioned experiments, it is
likely that heating of the reaction system or prolonged contact time of
approximately 3 weeks improves the recovery of radioruthenium, not only
by the adsorption on the manganese dioxide precipitate, but also by other
adsorption methods such as with ferric hydroxide, etc. Of course, the
large water samples are not readily heated.
    Another possible way to improve the recovery of radioruthenium may
be to go through an oxidation-reduction cycle of ruthenium, such as the
method used by Shiozaki [71] for 20-40 litres of seawater, though the procedure when applied to large-volume samples seems to be tedious. The
coprecipitation of radioruthenium with cobalt sulphide proposed by
Yamagata and Iwashima [72], on the other hand, proved to be an acceptable

procedure for γ-spectrometry with relatively small samples (5-10 litres) in which no carrier is added. Guéguéniat et al. [70] observed approximately 100% recovery of radioruthenium even for the effluents from the La Hague Plant (1972-1973) as well as those from the Marcoule Plant. The applicability of this method to large-volume samples has yet to be tested.

In conclusion, the important aspect of the application of the procedures without addition of stable carrier for monitoring or research purposes is to have a knowledge of the recovery of radioruthenium from the samples taken in the specific areas in question through these procedures, before they are applied to the in situ samples. Without this knowledge the results obtained are always under suspicion. In this respect, attention should be directed to the fact that the chemical forms of radioruthenium might change even from the same source, depending on changes in treatment as shown in the example cited above.

4. DETERMINATION ON BIOLOGICAL MATERIALS AND SEDIMENTS

Due to biological concentration processes the levels of radioruthenium in marine biota are usually high enough to allow the application of direct γ-spectrometry for its determination. Concentration factors of ruthenium among various types of marine organisms range normally between 1 and 2000. The levels of radioruthenium in marine sediments taken from a monitoring area are also high since they represent the sink for radioruthenium. For this reason there is a general tendency to use more and more direct γ-spectrometry for monitoring of radioruthenium in marine organisms and sediments. It seems, however, this method is not without problems.

Controversial opinions still exist about the loss of ruthenium during the ashing of marine organisms or the efficiency of dissolution of ruthenium from sediments by various procedures (radiochemical methods of determination). The points calling for special attention are discussed in this section.

4.1. Direct gamma spectrometry

As has been stated, the recent intercalibration exercises on radionuclides measurements in contaminated seaweed [9] and sediment [7] demonstrated a large scatter of the reported results, especially for [106]Ru, even though the majority of the laboratories employed non-destructive γ-spectrometry with Ge(Li) detectors for the determination of major γ-emitters. In the case of the sediment sample the percentage standard deviation of the average value for results obtained by Ge(Li) γ-spectrometry is ± 35%, which exceeds either that by NaI γ-spectrometry (± 9%) or that by radiochemical methods (± 15%). This indicates that there is a problem of calibration of Ge(Li) γ-spectrometers in some laboratories due, perhaps, to the different procedures adopted and/or the different standards used. In addition, the frequency distribution histograms of the reported results showed a subpeak somewhat lower than the main peak, which suggests that some measurements made by Ge(Li) or NaI γ-spectrometry might

involve systematic errors lowering the results obtained. This leads to
the suspicion that, in some cases, geometry factors or self-adsorption
corrections might not have been fully accounted for. Since direct γ-counting
on large samples, especially by Ge(Li) detectors, is sensitive to these
factors, care should be taken to offset these sources of errors in the calibration procedures.

In cases where the 512-keV γ-peak is used for the determination of
[106]Ru it should be kept in mind that any annihilation radiation produced by
positron emitters (511 keV) would be unresolvable from the [106]Rh peak,
giving systematically high results. The γ-peak of 622 keV is more suitable
for interference-free determination, even though the branching ratio for
this peak is smaller by a factor of 2 than that of the 512 keV peak.

4.2. Dry-ashing of marine biological samples

Concerning the loss of ruthenium during dry-ashing, different workers
have produced varying results. While Riley [73] observed that the loss of
[106]Ru from seaweed dry-ashed at 450-500°C did not exceed 5%, Strohal and
Jelisavcic [74] found only 89 and 70% of [106]Ru originally present in mussels
after drying at 110°C and ashing at 450°C, respectively. Iwashima [69]
followed the experimental procedures of Strohal and Jelisavcic, finding
that [106]Ru in mussels remained at 96% on drying 110°C and at 92% on
ashing at 450-500°C. Nakamara et al. [75] found, on the other hand, different losses of [106]Ru from seaweed and bivalues depending on varying
conditions of dry-ashing.

These results indicate that the loss of ruthenium during the dry-ashing
is not only dependent on the ashing conditions, but also on the species of
marine organisms treated. Moreover, it is quite likely that even in the
same species several factors such as different growth stages, the ways in
which ruthenium was incorporated into the organisms, the chemical states
of incorporated ruthenium in the body, etc. influence the loss of ruthenium
in ashing. For this reason, although recommended procedures in literature
may give sufficient recovery of ruthenium in the majority of cases, the loss
of radioruthenium from biological samples should always be carefully looked
at for each specific kind of sample, when destructive procedures are employed.

4.3. Dissolution from sediment samples

Although marine sediments represent a sink for radioruthenium released
into the marine environment, the mode and kinetics of incorporating radioruthenium into sediments under natural conditions are not well understood.
As has been discussed, however, various inorganic and organic complexes
and the polymerized forms of ruthenium are certainly involved in the incor-
poration processes. Since the chemical reactivity of these forms, especially
of polymerized forms, is not known, various dissolution procedures proposed to date are only empirical rather than logical. Nevertheless, the
alkali fusion method under oxidizing conditions after the destruction of
organic matter is considered as a sure method to dissolve ruthenium. In
this procedure, however, the effect of the destruction of organic matter by
ignition on possible volatile organic fractions of ruthenium has not been
well studied. The quantitative importance and behaviour of such organic
matter has yet to be examined.

Since the fusion method is so tedious in application to more than a few grams of sediment, various leaching procedures, especially with concentrated acids, have been employed by some workers [76, 77] to treat larger amounts of sediments. Fukai (private communication) found that two leachings of an air-dried sediment sample with aqua regia for 15 hours at room temperature and 5 hours at 90°C bring more than 99% of ^{106}Ru into solution, while more than 5% of ^{106}Ru remains in the same over-dried sediment after one leaching only. Guéguéniat (private communication) used 9$\underline{M}$ HCl for leaching on experimentally contaminated sediments with nitrato ruthenium complexes and obtained dissolution of 62% of ^{106}Ru from a sediment sample after a contact time of one hour and 92% after 20 hours. It seems that ^{106}Ru adsorbed by sediments in situ is effectively leached by two leachings with aqua regia, although the large amount of iron dissolved at the same time makes the succeeding procedures difficult when radiochemical separation methods are employed. In view of this, the perfection of distillation methods for the separation of ^{106}Ru in combination with the leaching procedure may be worth exploiting for treating large sediment samples. At present, these procedures seem to give systematically low results.

5.   GENERAL REMARKS

As has been fully discussed above, an accurate determination of radioruthenium in seawater, marine organisms and sediments depends very much on the knowledge of chemical forms of ruthenium in the various matrices of the samples. This knowledge is, unfortunately, far from complete. To fill in this gap in our knowledge, tests of some specific key analytical procedures by using radioactive tracers are still required in order to prove the correctness of the procedures adopted. In this connection it should be borne in mind that some commercial radioactive tracers, especially of radioruthenium, are very different from the chemical forms described and, sometimes, contain impurities, such as organic molecules, which seriously influence the results of experiments.

IV. ZIRCONIUM-95 (NIOBIUM-95)

1. INTRODUCTION

The current analytical position regarding ^{95}Zr/^{95}Nb has already been summarized in the review paper presented in this report [78]; this chapter is designed both to define areas where caution must be exercised, and to indicate where analytical techniques might be applied with some confidence.

2. SAMPLE COLLECTION AND STORAGE

The procedures and precautions mentioned in the previous Technical Report [5] should be adequate for the collection of most samples containing ^{95}Zr/^{95}Nb. An important point should be re-emphasized at this stage: this is the possibility of the contamination of biological materials (especially filter-feeding organisms and some seaweeds) by sediment, with which is associated a relatively greater proportion of ^{95}Zr/^{95}Nb activity. When sampling edible materials for control purposes it is necessary that the procedure used for the cleansing from silt prior to analysis should follow as closely as possible that used before eating the material. For ecological research purposes as much as possible of the free sediment and detritus should be removed from the sample by careful washing or immersing for some hours in filtered seawater from the same locality as the sample [79].

Jefferies (private communication) has indicated that for mud samples from the Cumberland coast, UK, the half-thickness value for ^{95}Zr/^{95}Nb penetration is about 1.2 cm. A 10 cm depth of core will therefore produce a satisfactory sample for investigation. In the case of sand samples a half-thickness of about 15 cm was obtained; the ratio of surface activity in the two types of sample is about 30:1.

For the storage of biological materials and sediments again no additional points are necessary. If, however, it is required to determine a total ^{95}Zr + ^{95}Nb, then the analysis should be carried out as soon as possible after collection because of the difficulty of interpreting the data due to the grow-in of the ^{95}Nb daughter [78]. The storage of seawater samples prior to analysis for ^{95}Zr/^{95}Nb may present problems. An investigation has been carried out into the adsorption of environmental ^{95}Zr from filtered seawater (using 0.45 μm pore diameter paper) onto polyethylene, polypropylene and polyvinyl chloride containers; the seawater was collected from the Irish Sea in the vicinity of the discharge from the British Nuclear Fuels Ltd factory at Windscale. The results of this investigation indicated that the ^{95}Zr that passed through the filter paper was not adsorbed on the container surfaces for at least up to a period of about 200 days from seawater both of natural pH and at pH 1.5-2 (by addition of hydrochloric acid). These findings may not be valid in other situations, where the ^{95}Zr may perhaps be in different physical or physico-chemical forms from those encountered in the Irish Sea, and similar experiments should be carried out if storage of samples from other environments is being considered.

3. PRETREATMENT AND CONCENTRATION

The treatment of biological materials prior to analysis presents few difficulties; there have been no reported losses of ^{95}Zr/^{95}Nb during drying and wet- and dry-ashing procedures, in the last instance at temperatures of up to 450°C, but it should be noted that strongly ignited ZrO_2 becomes refractory [80] and ignition temperatures should therefore be kept as low as possible. The dilemma concerning the inclusion or exclusion of associated siliceous material, previously mentioned in connection with sample collection, again faces the analyst during pretreatment; complete dissolution of biological materials (and also sediments) and removal of interfering silica can be achieved by treatment of the ashed residues with H_2SO_4/HF mixtures, followed by alkaline fusion.

The recommendations of the previous Technical Report [5] for the treatment of both biological materials and sediments should be carefully studied before carrying out any sample decomposition.

There are several well-proven techniques for the separation of ^{95}Zr/^{95}Nb from seawater: the use of manganese dioxide is the established technique at La Hague [81], and tests at the Fisheries Radiobiological Laboratory at Lowestoft have shown that this technique also quantitatively removes these radionuclides from seawater taken from the Irish Sea. The laboratory has also tested granular manganese dioxide [82] and metal-loaded anion and salt-loaded cation exchange columns [83] and found that all remove ^{95}Zr/^{95}Nb with virtually 100% efficiency. Hydroxide coprecipitations (using Fe^{III} and La^{III}) have also been shown to effect a quantitative removal of the radionuclides from filtered seawater [84] and other concentration procedures are listed in the review paper [78], but it would again be prudent for the analyst to check them carefully in the environment he is studying.

4. METHODS FOR DETERMINATION

4.1. Direct measurement

The review paper [78] discusses the sensitivity that might be required of an analytical method when considering the consumption of seafoods contaminated with ^{95}Zr/^{95}Nb; levels of 1-10 pCi/g wet material might be encountered in this context and at these levels some form of direct γ-spectrometric analysis of the dried sample is adequate. The use of either NaI(Tl) or Ge(Li) detectors for direct measurement of ^{95}Zr/^{95}Nb activity should also suffice for almost all research purposes. The presence of systematic errors in the interpretation of the spectra obtained can be checked in several ways, e.g.:

(a) By intercomparison using both types of detector to measure the activity of the same sample

(b) By the measurement, at intervals over several months, of an artificial standard containing other radionuclides typically present in the samples, together with equilibrated ^{95}Zr/^{95}Nb (activity ratio 1:2.17). The activity level of ^{95}Zr/^{95}Nb at which deviation from the decay of ^{95}Zr occurs indicates when interpretive errors become significant in the samples.

Careful investigation of the spectra produced by the Ge(Li) detector in method (a) should indicate whether there are any radionuclides that might interfere in the spectral unfolding technique used for determination of the activity from the NaI(Tl) spectra.

4.2. Chemical separations

If very low levels of ^{95}Zr are encountered or if interference from other radionuclides makes the satisfactory interpretation of the gamma spectrum impossible, then chemical separations may have to be applied. These can be quite satisfactory, provided it is remembered that zirconium ions readily undergo hydrolytic polymerization and form colloids.

Separations schemes have been reported using carrier addition and isotopic equilibration with oxalic acid [85] and hydrofluoric acid [86]; the latter reagent is particularly useful because it forms the very stable ZrF_6^{2-} complex; both ensure the dissolution of the very insoluble zirconium phosphate. Subsequent separations include the precipitation of barium fluorzirconate and anion-exchange separation in HCl/HF media in the two methods mentioned.

Several other schemes for the separation of ^{95}Zr or ^{95}Zr + ^{95}Nb from seawater and seaweeds are referred to in the review papers (e.g. Refs [87-89]). Methods that do not employ carrier exchange or use carrier-free stages should be carefully checked before application, but it does seem that the propensity for zirconium to be easily coprecipitated and for radiozirconium to be adequately stabilized in the presence of HF or strong organic complexing agents such as tributylphosphate or thenoyltrifluoracetone can ensure high and reproducible yields.

4.3. Source preparation and measurement of activity

Very low levels of activity must be measured on a low background ($\approx$ 1 counts/min) beta counter, while higher levels can also be measured by gamma spectrometry. For this latter method the ^{95}Zr/^{95}Nb fraction from any suitable stage in the separation can be counted in a convenient geometry. If beta counting is necessary, then care must be taken to produce a thin, reproducible source because the low energy of the beta-emissions from ^{95}Zr − and the even lower energy of those from its daughter ^{95}Nb − result in high self-absorption of the source. Examples of sources used are ZrO_2 (produced by ignition) and zirconium mandelate. In the procedure reported by Ikeda et al. [84], the organic extract can be evaporated to dryness for counting; only 5 μg of zirconium (carrier) are used in this technique and the resultant source is virtually free from self-absorption. If the sources are thick, then a graph of counter efficiency versus weight of precipitate on the tray should be prepared using aliquots of standardized ^{95}Zr/^{95}Nb solution.

The gel-scintillation counting of the separated (white) zirconium sources can also be considered, with the possibility of discrimination against the lower energy β^- emissions of ^{95}Nb.

5. DATA INTERPRETATION

The difficulties associated with the interpretation of ^{95}Zr/^{95}Nb results obtained by NaI(Tl) spectrometry are referred to in the review paper [78]. These problems are generally overcome by the use of Ge(Li) spectrometry, which can adequately distinguish between ^{95}Zr and ^{95}Nb.

6. GENERAL REMARKS

Some form of gamma spectrometry should generally produce adequately precise and accurate measurement of ^{95}Zr/^{95}Nb for both control and research purposes, provided adequate care is given to the interpretation of the spectra. Beta or gamma counting of the chemically separated fractions may infrequently be necessary to determine ^{95}Zr for research purposes; adequate methods are available, provided they are carefully tested for the particular environment under investigation.

V. RADIOACTIVE AND STABLE SILVER

1. INTRODUCTION

Reported methods for the determination of radioactive and stable silver in the marine environment have been covered by the review paper presented in this report [90]. This chapter is intended to complete the analytical picture and to indicate areas where further development is required.

2. SAMPLING, STORAGE AND PRETREATMENT

Obviously, any sampling procedure should be checked for reliability, but reported levels of stable silver in standard materials such as polyethylene and polytetrafluorethylene are low [91] and their use should not lead to contamination of seawater, biological materials or sediments.

The storage of seawater samples should again present no difficulties: Schutz and Turekian [92] reported no significant adsorption of silver into Pyrex over a period of 6 months, and although Robertson [93] found that 20% of the silver was lost from seawater stored in Polythene after only 10 days, this loss was prevented by the acidification of the sample to pH 1.5 with hydrochloric acid. This reagent is available in a form pure enough to ensure no significant contamination of the sample. Table VIII shows losses of silver onto polyethylene from unfiltered and filtered (through 0.22 μm diameter Millipore filter papers) seawaters under various conditions [97].

These data indicate that, although freezing (to -20°C) and acidification to pH 2 are both suitable for storing seawater for periods of up to 28 days, acidification is perhaps preferable.

Losses of added silver during the ashing of biological materials have been reported. During investigations into dry-ashing techniques, Gorsuch [95] found that at about 550°C recoveries of silver added to the sample varied between 87 and 100%, depending on the ashing aid used, the greatest losses occurring when nitric acid was employed, with 7% retained on the silica vessel. Pijck et al. [96] used blood spiked with $^{110}Ag^m$ and found that at 400°C 35% was lost in 24 hours, and at 900°C the loss of silver amounted to 79% in 3 hours. Both Gorsuch [95] and Pijck et al. [96] found that recoveries of silver during wet-ashing procedures with nitric, perchloric and sulphuric acid mixtures were > 95%. Gleit and Holland [97] reported that low-temperature ashing (using R.F. excited oxygen) resulted in a 27% loss of silver after 2 hours; losses of silver have also been reported by Hamilton et al. [98] when ashing using this technique. No losses during the oven drying (at 105°C) of biological materials have been noted by UK workers (Dutton, private communication).

In the light of the above summary, wet-ashing techniques are recommended if the destruction of organic material is necessary prior to the analysis of silver. Careful investigations into the various dry-ashing techniques, using organisms environmentally labelled with $^{110}Ag^m$, should result in the production of much valuable information. In the meanwhile, great care should be taken to check any methods used.

TABLE VIII. LOSSES OF SILVER FROM SEAWATER ONTO
POLYETHYLENE [94]

Storage conditions			Losses (%)				
			1 h	1 d	7 d	28 d	
pH 8	light,	20°C	3	9	43	79	Unfiltered seawater
	dark,	20°C	1	-	13	81	
	dark,	-20°C	-	4	3	7	
pH 2	light,	20°C	3	2	< 1	3	Unfiltered seawater
	dark,	20°C	3	1	< 1	3	
pH 8	light,	20°C	1	< 1	6	25	Filtered seawater
	dark,	20°C	1	< 1	8	46	
	dark,	-20°C	-	< 1	< 1	3	
pH 2	light,	20°C	< 1	< 1	< 1	< 1	Filtered seawater
	dark,	20°C	< 1	< 1	< 1	< 1	

3.   DETERMINATION OF RADIOSILVER

The review paper [90] contains details of the sensitivity of various
instrumental methods used for the direct determination of both $^{110}Ag^m$
and $^{108}Ag^m$, and intending analysts should be certain of their requirements
before choosing an analytical system. In a monitoring context, levels of
$^{110}Ag^m$ considerably lower than 1% of the derived working limit (calculated
from ICRP recommendations) can be measured using direct NaI(Tl) gamma
spectrometry. When direct instrumental methods are chosen – and hence
interpretative techniques are used – it is highly desirable to carry out
independent analyses to check the validity of the method, and the importance
of quality control [50] cannot be overemphasized.

The recent intercomparison exchange carried out by IAEA involving
environmentally labelled Fucus seaweed resulted in several laboratories
reporting data on $^{110}Ag^m$ [9], but due to the paucity of data and the low $^{110}Ag^m$
level, no firm conclusions can be drawn from this exercise; no other
collaborative studies of analytical methods for $^{110}Ag^m$ have been reported
and work in this field should be encouraged – it might be possible to
ensure comparability of analytical methods before $^{110}Ag^m$ data are
extensively reported.

Reports of the contamination of analytical materials by $^{110}Ag^m$ have
been limited to one only – the discovery of its presence in lead shielding [99] –
but it follows from the increased use of radionuclides and radiation in
technology that analysts should look very carefully at their reagents and
apparatus.

4. DETERMINATION OF STABLE SILVER

Methods for the determination of stable silver in marine environmental materials have been summarized in the review paper [90]. A further method for seawater analysis [100] not discussed in that paper involves preconcentration by coprecipitation with lead sulphide followed by dithizone separation and spectrophotometric measurement — $^{110}Ag^m$ is used to determine the chemical yield; good agreement was obtained with an independent neutron activity analysis. Attention has already been drawn [90] to the discrepancies in reported results for stable silver in a kale intercomparison sample [101]. In view of the interest shown in silver in ecology, this problem should be resolved as speedily as possible.

5. GENERAL REMARKS

Gamma spectrometry in some form is the obvious technique for the measurement of both $^{110}Ag^m$ and $^{108}Ag^m$ in the marine environment for both control and research purposes. The intercomparison exercises involving methods for the determination of both radioactive and stable silver are strongly urged.

VI. RADIOIODINE

1. IODINE-131

1.1. General

The presence of ^{131}I in the marine environment from stratospheric fall-out has not been reported, and the techniques discussed here are designed to measure ^{131}I and, in some cases, other iodine radioisotopes resulting from discharges from nuclear establishments. Derived working levels for ^{131}I in some seafoods — based on ICRP recommendations and possible consumption rates — are given in the review paper presented by Dutton [102], together with 1% of these levels; these are considered suitable limits of detection to serve as a basis for the assessment of analytical techniques.

Standard methods for the determination of ^{131}I have been published [103] and these methods, together with others including those presented in this report, are discussed below.

1.2. Sample collection and storage

The general precautions for collection and storage of seawater samples presented in the previous Technical Report [5] should be followed for ^{131}I. The problems associated with long-term storage of samples will not arise because of the short half-life of the radionuclide (8.04 days) and for this reason alone the analysis should be carried out as soon as possible after collection of the sample. No losses of ^{131}I were reported from filtered seawater stored, unacidified, for a few days in polyethylene bottles by the two laboratories carrying out the determinations of ^{131}I [102; Pillai, private communication], but it is important that checks on any possible losses of the radionuclide during storage be carried out: any I_2 present may be lost either by volatilization or by dissolution in the walls of the plastic containers. Losses of iodine from biological materials, especially certain species of algae, may occur during their transport to the laboratory; it is therefore recommended that such samples are frozen, or at least kept cool, during this time.

1.3. Sample pretreatment and concentration

Although it has been reported that ^{131}I has not been lost during the oven-drying (at 100-110°C) of <u>Fucus</u> seaweed [102], it should not be assumed that this will be the case for all other biological materials. Gamma-spectrometric analysis can be applied to the original wet sample [102], which should, of course, be properly homogenized.

If chemical separation of ^{131}I from biological materials is required, then the organic matrix must be destroyed. Acid oxidations, using chromic acid [104] or potassium permanganate and sulphuric acid [103, 105], have been employed to achieve this. In addition, alkaline oxidative destruction has been used, involving mineralization of fusion with sodium hydroxide followed by fusion with potassium nitrate [106]. It appears that the alkaline

fusion technique is suitable for larger quantities of organic matter. Direct
dry-ashing of materials should not be countenanced, and even low-temperature
ashing using oxygen excited by a radiofrequency discharge has resulted in
losses of up to 80% of ^{131}I from rat tissues [98].

1.4. Direct gamma spectrometry

The use of NaI(Tl) gamma spectrometry has been reported for the
measurement of ^{131}I in <u>Fucus</u> seaweed; with a sample size of 2 kg (wet)
and a counting time of 60 minutes, levels of ^{131}I down to 0.05 pCi/g
(i.e. a total of 100 pCi in the sample) were easily measured [102]. The
possible presence of interfering radionuclides must be kept in mind when
using NaI(Tl) detection — for example, [140]La (daughter of [140]Ba), which
emits a gamma photon at 0.33 MeV — and if such interference is suspected,
then recourse must be made either to Ge(Li) detection or chemical
separation.

Systematic errors may occur due to incorrect interpretation of
gamma-spectral data: the decay of ^{131}I should be followed by recounting
at intervals to check that these errors are not significant, and the measurement of ^{131}I in selected samples may also be carried out by chemical
methods to check the validity of the interpretative technique.

1.5. Chemical separations

Although direct gamma-spectrometric analysis is sufficiently sensitive
to allow proper control of ^{131}I disposals to the marine environment, interferences from other radionuclides may reduce the quality of data sufficiently
to warrant chemical separations of the radionuclide. In addition, radioecological problems may require the analysis of materials (including biota
and seawater) where the level of ^{131}I is low enough to make gamma spectrometry impracticable. In all these cases separation of the radionuclide
is necessary.

1.5.1. Isotopic equilibration

It is general radiochemical practice to ensure that the radioactive
iodine in the sample and the added carrier are in the same chemical form,
and this is often achieved in liquids — including seawater — by the addition
of carrier as iodide (I^-), oxidation to periodate (IO_4^-) with NaOCl in
alkaline medium, followed by reduction to I^- again with hydroxylamine or
bisulphite in acid solution [107]. Both I^- and IO_3^- (iodate) forms have been
reported in seawater [108] and if it is proposed to shorten or omit alto-
gether the oxidation/reduction stages, then it would be advisable to test
the method both with and without the complete exchange cycle. For biological materials isotopic equilibration appears to be achieved by the
oxidative processes used to achieve destruction of the organic matrix
(section 1.3), followed by reduction to free iodine with oxalic acid [103].

1.5.2. Separation from other radionuclides

The two procedures generally used to separate ^{131}I from other radionuclides are the distillation and the solvent extraction of free iodine [102, 105].

Distillation has, however, been found to be impracticable for amounts of biological materials above 20-30 grams [106] and it is also difficult to use the procedure with the relatively large volumes (about 5 litres) of seawater; solvent extraction is therefore suggested as a suitable technique. Carbon tetrachloride is the preferred organic solvent for the extraction of iodine, which is liberated by oxidation of the iodide using sodium nitrite and nitric acid. Back extraction into the aqueous phase is achieved by the use of sulphur dioxide solution or sodium bisulphite and acid.

Recently, two methods have been proposed using exchange techniques, on silver chloride [109] and silver iodide [110], to achieve separation of ^{131}I from seawater and this technique is very attractive. Decontamination factors from most radionuclides other than silver are very high, although some adsorption of ^{95}Zr(-^{95}Nb) on the silver iodide column was noted. Separation of silver iodide (together with the bromide naturally present in the seawater) from the silver chloride column is effected by treatment with ammonia solution; the iodide is then dissolved in potassium cyanide solution and reprecipitated by acidification with nitric acid. [CAUTION — cyanide solutions and HCN (evolved on acidification) are extremely poisonous.] This precipitate is freed from the large amount of bromide by heating on a water bath with potassium dichromate and sulphuric acid, thus forming free bromine. Some losses may occur at this stage due to the formation of free iodine, and it is recommended that great care be taken during this oxidation step.

1.6. Counting methods and source preparation

The choice of a counting technique depends on the amount of radioactivity available. For very low activities use must be made of beta-counting, with, for example, a 50 mm diameter anti-coincidence-shielded gas-flow detector after chemical separation of the radioiodine; backgrounds of 1 count/min with counting efficiencies of 40% can be obtained for beta-counting equipment, and the determination of 1 pCi of ^{131}I is therefore easily achieved.

Suitable sources for beta counting can be made of thinly spread precipitates of silver iodide or palladous iodide, but care should be taken when drying the latter compound because it is easily decomposed with loss of iodine in vacuo or by warming strongly. Normal desiccation over silica gel, however, is satisfactory.

For higher levels of activity, simple gamma counting (gated to cover the 0.364 MeV peak) can be used to measure the separated source. Alternatively, gamma-spectrometric measurement can be carried out on a partially purified iodine fraction after concentration by the various methods mentioned earlier. An advantage of the isotopic exchange method using silver iodide is that a gamma-spectrometric analysis can be carried out after this separation step.

2. OTHER RADIOISOTOPES OF IODINE

If very short-cooled fission products are discharged into the marine environment, then ^{132}I, ^{133}I and ^{135}I may be encountered. In this case the

TABLE IX. DECAY CHARACTERISTICS OF IODINE RADIONUCLIDES

	^{129}I	^{131}I	^{132}I a	^{133}I	^{135}I
Half-life	1.6×10^7 a	8.04 d	2.3 h	20.8 h	6.7 h
β_{max} (MeV)	0.150 (100%)	0.608 (87%) others	0.73-2.12	1.4 ($\approx$94%) others	1.5 others
γ (MeV)	0.038	0.364 (81%) others	0.673 (100%) 0.777 (85%)	0.53 ($\approx$94%) others	1.14, 1.27 others
Fission yield (%)	0.9	2.9	4.3	6.5	7.9

a Daughter of ^{132}Te, half-life 77.7 h.

use of gamma spectrometry would be essential, and prior chemical separation may be necessary if other interfering radionuclides do not permit a satisfactory interpretation of the spectra.

A very long-lived fission product, ^{129}I (half-life 1.6 × 10^7 years), has been the subject of attention as an ecological and geochemical tracer [111] and as a possible radiological hazard [111, 112]. Methods for the determination of this radionuclide are being developed [113]. The separation techniques will be similar to those mentioned above, but the measurement of ^{129}I may cause some difficulty on account if its low specific activity in the marine environment.

The decay characteristics of the radionuclides of iodine discussed above are given in Table IX, in which ^{131}I has also been included.

3.   GENERAL REMARKS

Standard gamma-spectrometric procedures and reported chemical separations appear to be adequate for the measurement of ^{131}I in the marine environment. Unfortunately, few data on the intercomparability of methods have been reported; any such data − e.g. the comparison of gamma-spectrometric with chemical-separated ^{131}I results and the comparison of standard procedures with the exchange column methods of determining ^{131}I in seawater − should be widely circulated. There is also little data on the storage stability of ^{131}I in seawater and its losses during the pretreatment of marine biological materials. In all these cases more programmes of investigation are required and they should be carried out in environments where radioiodine contamination occurs.

VII. GENERAL RECOMMENDATIONS

The general recommendations agreed upon at the panel, especially for the International Atomic Energy Agency, were as follows:

(1) Publish the work of the Panel as an Agency Technical Report, "Reference Methods for Marine Radioactivity Studies, II Analytical methods for transuranic elements, ruthenium, zirconium, silver and iodine". This would be an adjunct to Technical Report 118.

(2) Consider the requirement to update the work of the present panel and the panel that produced Technical Report 118 at some time in the future. In addition, there may be a future requirement for reference methods for other radionuclides not covered by the two panels.

(3) Sponsor Panels on Transport, Distribution and Behaviour of Transuranic Elements in the Aquatic Environment and Distribution and Behaviour of Ruthenium in the Aquatic Environment. These might be held in 2 or 3 years and include, where possible, studies on the physical and chemical states in the environment of the elements involved.

(4) Encourage the expansion of capacity to measure transuranic isotope ratios in low-level environmental samples.[1]

(5) Encourage research or development in fields where the panel found specific needs, particularly:
  (a) Methods for analysis of transplutonium elements and antimony-125, and reliable control-type methods for plutonium, radioiodine, radioruthenium and radiozirconium;
  (b) Methods to determine the physical and physico-chemical state of the transuranic elements, ruthenium and zirconium in the environment;
  (c) Studies of the transport, distribution and state of these elements in coastal and estuarine waters, since these waters are most commonly used for waste disposal and also are used most intensively by man;
  (d) Production of reliable instruments of moderate sophistication for alpha spectrometry.

(6) Include the following in the programme of the Monaco Laboratory:
  (a) Continue the highly successful programme of preparation and distribution of intercomparison samples of environmental material, both marine and freshwater. These samples should continue to be of materials environmentally contaminated with either fission-produced or activation-produced nuclides; participants in these intercalibrations should be urged to measure as many constituents as they can find, especially transuranic nuclides, ruthenium, zirconium and silver;
  (b) Continue the present programme of co-operation with the Seibersdorf Laboratory in the preparation and distribution of standard reference samples. This is a service of primary importance in environmental studies;

---

[1] The Lawrence Livermore Laboratory in the United States of America has the required isotope separator and its capacity could be expanded at relatively modest cost.

(c) In co-operation with the Seibersdorf Laboratory, distribute
calibrated solutions suitable for yield monitoring and standardization in the analysis for transuranic elements;

(d) Maintain an analytical capability for the purpose of testing
intercomparison samples and reference samples and for
effective evaluation of these programmes. In particular, it
would be helpful to determine the reasons for the present
discrepancies in ruthenium analysis;

(e) Develop an analytical capability for the spectrometric
measurement of alpha-emitting radionuclides;

(f) Since the activities described require a considerable degree
of competence in both the analytical chemistry and the understanding of the ecological processes involved, it is necessary
to attract suitable staff. Operation of a modest research
programme relevant to the fields described would help to make
the staff positions attractive.

REFERENCES

[1] UNITED NATIONS, Document A./CONF. 13/38 (1958) 143-44.

[2] INTERNATIONAL ATOMIC ENERGY AGENCY, Radioactive Waste Disposal into the Sea, Safety Series No. 5, IAEA, Vienna (1961) 174 pp.

[3] INTERNATIONAL ATOMIC ENERGY AGENCY, Methods of Surveying and Monitoring Marine Radioactivity, Safety Series No. 11, IAEA, Vienna (1965) 95 pp.

[4] SLANSKY, C.M., Principles for limiting the introduction of radioactive waste into the sea, At. Energy Rev. 9 (1971) 853-68.

[5] INTERNATIONAL ATOMIC ENERGY AGENCY, Reference Methods for Marine Radioactivity Studies, Technical Reports Series No. 118, IAEA, Vienna (1970) 284 pp.

[6] FUKAI, R., BALLESTRA, S., MURRAY, C.N., "Intercalibration of methods for measuring fission products in sea water samples", Radioactive Contamination of the Marine Environment (Proc. Symp. Seattle, 1972), IAEA, Vienna (1973) 3-27.

[7] FUKAI, R., STATHAM, G., BALLESTRA, S., ASARI, K., "Intercalibration of methods for radio-nuclide measurements on a marine sediment sample", Environmental Surveillance around Nuclear Installations (Proc. Symp. Warsaw, 1973), IAEA, Vienna (1974) 313-35.

[8] FUKAI, R., MURRAY, C.N., "Results of plutonium intercalibration in seawater and seaweed samples", this report.

[9] FUKAI, R., BALLESTRA, S., "Intercalibration of methods for radionuclide measurements on a seaweed sample", this report.

[10] ANONYMOUS, Project Crested Ice, Risø Rep. 213 (1970).

[11] JOHNSON, L.J., Los Alamos Land Areas Environmental Radiation Survey, Rep. LA-5097-MS (1972) 27 pp.

[12] BALLOU, J.E., THOMSON, R.C., CLARKE, W.J., PALOTAY, J.L., Comparative toxicity of plutonium-238 and plutonium-239 in the rat, Health Phys. 13 (1967) 1087-92.

[13] MORIN, M., NENOT, J.C., LAFUMA, J., Metabolic and therapeutic study following administration to rats of ^{238}Pu nitrate: A comparison with ^{239}Pu, Health Phys. 23 (1972) 475-80.

[14] HAKONSON, T.E., JOHNSON, L.J., Distribution of Environmental Plutonium in the Trinity Site Ecosystem after 27 Years, Rep. Conf. 730907-14, LA-UR-73-1291 (1973) 6 pp.

[15] RAABE, O.G., KANAPILLY, G.M., BOYD, H.A., Studies of the in vivo Solubility of ^{238}Pu and ^{239}Pu Oxides and Accidentally Released Aerosol Containing ^{239}Pu, Lovelace Foundation Ann. Rep. 1972-1973, LF-46, UC-48 (1973) 24-30.

[16] PRICE, K.R., A review of transuranic elements in soils, plants and animals, J. Environm. Qual. 2 (1973) 62-66.

[17] NOSHKIN, V.E., BOWEN, V.T., "Concentrations and distribution of long-lived fallout radionuclides in open ocean sediments", Radioactive Contamination of the Marine Environment (Proc. Symp. Seattle, 1972), IAEA, Vienna (1973) 671-86.

[18] BURKE, J.C., A sediment coring device of 21-cm diameter with sphincter core retainer, Limnol. Oceanogr. 13 (1968) 714-78.

[19] AARKROG, A., "Radiochemical determination of plutonium in marine samples by ion exchange and solvent extraction", this report.

[20] WONG, K.M., NOSHKIN, V.E., BOWEN, V.T., "Radiochemical procedures of Wong (now used at WHOI) for the analysis of strontium, antimony, rare earths, cesium and plutonium in sea water samples", Reference Methods for Marine Radioactivity Studies, Technical Reports Series No. 118, IAEA, Vienna (1970) 119-27.

[21] DUTTON, J.W.R., IBETT, R.D., WOOLNER, L.E., "A study of the stability of some radionuclides in seawater", this report.

[22] BOJANOWSKI, R., LIVINGSTON, H.D., SCHNEIDER, D.L., MANN, D.R., "A procedure for analysis of americium in marine environmental samples", this report.

[23] SUGIHARA, T.T., JAMES, H.I., TROIANELLO, E.J., BOWEN, V.T., Radiochemical separation of fission products from large volumes of sea water: Strontium, cesium, cerium and promethium, Anal. Chem. 31 (1959) 44-49.

[24] INTERNATIONAL ATOMIC ENERGY AGENCY, International Laboratory of Marine Radioactivity, Monaco, Progress Rep. No. 10 on the Intercalibration of Analytical Methods on Marine Environmental Samples (1974) 4 pp.

[25] HAMPSON, B.L., TENNANT, D., Simultaneous determination of actinide nuclides in environmental materials by solvent extraction and alpha spectrometry, Analyst 98 (1973) 873-85.

[26] HARLEY, J.H., USAEC Health and Safety Laboratory Manual of Methods (1972).

[27] SURLS, J. P., CHOPPIN, G. R., Ion-exchange study of thiocyanate complexes of the actinides and lanthanides, J. Inorg. Nucl. Chem. 4 (1957) 62-73.

[28] SCHNEIDER, K. J., Advance Waste Management Studies – High Level Waste Alternatives, Doc. BNWL-1900 (1974).

[29] BUDNITZ, R. J., Plutonium, a Review of Measurement Techniques for Environmental Monitoring, IEEE (Inst. Electr. Electron. Eng.) Trans. Nucl. Sci. 21 (1973) 430-37.

[30] THOMAS, C. W., REID, D. L., LUST, L. F., Radiochemical Analysis of Marine Biological Samples Following the "Redwing" Shot Series, Rep. HW-58674 (1958) 82 pp.

[31] PILLAI, K. C., SMITH, R. C., FOLSOM, T. R., Plutonium in the marine environment, Nature (London) 203 (1964) 568-71.

[32] TEMPLETON, W. L., PRESTON, A., "Transport and distribution of radioactive effluents in coastal and estuarine waters of the United Kingdom", Disposal of Radioactive Wastes into Seas, Oceans and Surface Waters (Proc. Symp. Vienna, 1966), IAEA, Vienna (1966) 267-89.

[33] WONG, K. M., Radiochemical determination of plutonium in sea water, sediments and marine organisms, Anal. Chim. Acta 56 (1971) 355-64.

[34] LIVINGSTON, H. D., MANN, D. R., BOWEN, V. T., "Double-tracer studies to optimize conditions for the radiochemical separation of plutonium from large-volume seawater samples", this report.

[35] PILLAI, K. C., "Determination of plutonium in the marine environment", this report.

[36] LIVINGSTON, H. D., BOWEN, V. T., MANN, D. R., "Analytical procedures for transuranic elements in sea water and marine sediments", Am. Chem. Soc. Symp. Anal. Methods in Oceanography, 1974.

[37] MARKUSSEN, E. K., Danish Atomic Energy Commission, Risø-M-1242, unpubl. ms (1970).

[38] TALVITIE, N. A., Electrodeposition of actinides for alpha spectrometric determination, Anal. Chem. 44 (1972) 280-89.

[39] MITCHELL, R. F., Electrodeposition of actinide elements at tracer concentrations, Anal. Chem. (1960) 326.

[40] DARRALL, K. G., HAMMOND, G. C. M., TYLER, J. F. C., The determination of plutonium-241 in effluents, Analyst 98 (1973) 358-63.

[41] KEOUGH, R. F., POWERS, G. J., Determination of plutonium in biological materials by extraction and liquid scintillation counting, Anal. Chem. 42 (1970) 419-21.

[42] SAKANOUE, M., NAKAMURA, M., IMAI, T., "Determination of plutonium in environmental samples", Rapid Methods for Measuring Radioactivity in the Environment (Proc. Symp. Neuherberg, 1971), IAEA, Vienna (1971) 171-81.

[43] SILL, C. W., OLSON, D. G., Sources and prevention of recoil contamination of solid-state alpha detectors, Anal. Chem. 42 (1970) 1596-1607.

[44] CHETHAM-STRODE, A., TARRANT, J. R., SILVA, R. J., The application of silicon detectors to alpha particle spectroscopy, IEEE (Inst. Electr. Electron. Eng.) Trans. Nucl. Sci. 8 (1961) 59-63.

[45] KREY, P. W., KRAJEWSKI, B. T., Plutonium Isotope Ratio at Rocky Flats, Rep. HASL-249, I-67 to I-94 (1972).

[46] HARDY, E. P., KREY, P. W., VOLCHOK, H. L., Plutonium Fallout in Utah, Rep. HASL-257, I-95 to I-118 (1972).

[47] NOSHKIN, V. E., GATROUSIS, C., Fallout plutonium-240 and plutonium-239 in Atlantic marine samples, Earth Planet. Sci. Lett. 22 (1974) 111-17.

[48] LIVINGSTON, H. D., SCHNEIDER, D. L., BOWEN, V. T., Plutonium-241 in the marine environment by a radiochemical procedure, Earth Planet. Sci. Lett. (in press).

[49] DIAMOND, H., et al., Heavy isotope abundance in MIKE thermonuclear devices, Phys. Rev. 119 (1960) 2000-4.

[50] HARLEY, J. H., VOLCHOK, H. L., BOWEN, V. T., "Quality control in radiochemical analysis", this report.

[51] BITTEL, R., Discussion bibliographique sur le comportement physico-chimique et la radioécologie du ruthénium dans les systèmes hydrobiologiques, CEA BIB-123, EUR-3863 F (1968).

[52] POMAROLA, J., FELIERS, P., TESTEMALE, G., "Evolution de la gestion des déchets radioactifs au Centre de Fontenay-aux-Roses", Management of Low- and Intermediate-Level Radioactive Waters (Proc. Symp. Aix-en-Provence, 1970), IAEA, Vienna (1970) 537-62.

[53] MAGNO, P. J., "Studies of dose pathways from a nuclear fuel reprocessing plant", Environmental Behaviour of Radionuclides Released in the Nuclear Industry (Proc. Symp. Aix-en-Provence, 1973), IAEA, Vienna (1973) 537-51.

[54] MITCHELL, N. T., Radioactivity in surface and coastal waters of the British Isles 1971, Fisheries Radiobiological Laboratory, Lowestoft, Tech. Rep. FRL9 (1973) 31 pp.

[55] PILLAI, K. C., DEY, N. N., "Radioruthenium in coastal waters", this report.

[56] SHIOZAKI, M., YAMAGATA, N., IWASHIMA, K., "The analytical methods of ruthenium-106 in marine environmental samples (review paper)", this report.

[57] FLETCHER, J. M., BROWN, P. G. M., GARDNER, E. R., HARDY, C. J., WAIN, A. G., WOODHEAD, J. L., J. Inorg. Nucl. Chem. 12 (1959) 154-73.

[58] STORY, A. H., GLOYNA, E. F., Radioactivity Transport in Water. Environmental Behaviour of Nitrosyl-Ruthenium, Rep. ORO-620 (1963).

[59] BHAGAT, S. K., GLOYNA, E. F., Transport of Nitrosyl-Ruthenium in an Aquatic Environment, Tech. Rep. to AEC, EHE-11-6502 (TID-22-870) (1965).

[60] FLETCHER, J. M., JENKINS, I. L., LEVER, F. M., MARTIN, F. S., POWELL, A. R., TODD, R., Nitrato and nitro complexes of nitrosyl-ruthenium, J. Inorg. Nucl. Chem. 1 (1955) 378-401.

[61] BROWN, P. G. M., FLETCHER, J. M., WAIN, A. G., Nitrato Nitrosyl-Ruthenium and Their Extraction from Nitric Acid Systems by Tributyl Phosphate, Rep. AERE-C/R-2260 (1957).

[62] WAIN, A. G., BROWN, P. G. M., FLETCHER, J. M., The chromatographic separation of nitrato complexes of nitrosyl-ruthenium in nitric acid solution, Chem. Ind. (London) (5 Jan. 1957).

[63] VAN RAAPHORST, J. G., DEURLOO, P. A., The Cationic Species of Ruthenium Nitrosyl-Nitrato Complexes, Rep. ER. 52 (1963).

[64] GUÉGUÉNIAT, P., Nouvelles études sur les formes insolubles et sur les phénomènes de polymérisation des nitratocomplexes du nitrosylruthénium en eau de mer, C. R. Acad. Sci., Paris 274 (1972) 822-25.

[65] CHESSELET, R., LALOU, C., NORDEMANN, D., "Méthodes et résultats de mesure de radioactivité dans la mer", Symp. Contamination of the Marine Environment, C. E. R. B. O. M., Nice (1964).

[66] LAPICQUE, G., "Mesures "in situ" de l'activité de radioéléments dans le site marin d'une usine atomique et application au calcul du coefficient de diffusion turbulente", 23rd Congrès-Assemblée plénière de la C. I. E. S. M., Athens (Nov. 1972).

[67] LOVERIDGE, B. A., THOMAS, A. M., The Determination of Radioruthenium in Effluent and Sea Water, Rep. AERE C/R 2828 (1959).

[68] WORLD HEALTH ORGANIZATION, Methods of Radiochemical Analysis, WHO, Geneva (1966) 84-93.

[69] IWASHIMA, K., Analytical methods for ruthenium-106 in marine samples, J. Radiat. Res. 13 (1972) 127-48.

[70] GUÉGUÉNIAT, P., GANDON, R., LUCAS, Y., "Determination of radionuclides of cerium, cobalt, iron, ruthenium, zinc and zirconium in seawater by preconcentration with colloidal manganese dioxide. Application to the determination of low-level ruthenium-106", this report.

[71] SHIOZAKI, M., Ruthenium-106 in the adjacent sea of Japan, Rep. Hydrogr. Res. 1 (1966) 33-35.

[72] YAMAGATA, N., IWASHIMA, K., "Cobalt sulphides, an effective collector for radioruthenium complexes in sea-water", Rapid Methods for Measuring Radioactivity in the Environment (Proc. Symp. Neuherberg, 1971), IAEA, Vienna (1971) 85-90.

[73] RILEY, C. J., The Fate of Ruthenium-106 on Ignition of Seaweed (Porphyra) During Analysis, PG Rep. No. 122 (1962).

[74] STROHAL, P., JELISAVCIC, O., The loss of cerium, cobalt, manganese, protactinium, ruthenium, and zinc during dry ashing of biological material, Analyst 94 (1969) 678-80.

[75] NAKAMURA, R., SUZUKI, Y., UEDA, T., The loss of radionuclides in marine organisms during thermal decomposition, J. Radiat. Res. 13 (1972) 149-55.

[76] SHIOZAKI, M., ODA, K., KIMURA, T., SETO, Y., "The artificial radioactivity in sea water", Researches in Hydrography and Oceanography, Commem. Publ. Centenary of the Hydrographic Department of Japan, Tokyo (1972) 203-49.

[77] KAUTSKY, H., "Possible accumulation of discrete radioactive elements in river mouths", Disposal of Radioactive Wastes into Seas, Oceans and Surface Waters (Proc. Symp. Vienna, 1966), IAEA, Vienna (1966) 163-75.

[78] DUTTON, J. W. R., "A review of methods for the determination of radioactive and stable zirconium in the marine environment", this report.

[79] PATEL, B., BHATTATHIRI, P. M. A., DOSHI, G. R., Uptake of manganese-54 by ark-shell mollusc Anadara granosa (Linn.), J. Anim. Morphol. Physiol. 13 (1966) 158-68.

[80] ERDEY, L., Gravimetric Analysis, Part 2, Pergamon Press, Oxford, New York, London, Paris (1965) 474.

[81] GUÉGUÉNIAT, P., Détermination de la radioactivité de l'eau de mer en ruthénium, cérium, zirconium par entraînement et adsorption au moyen du bioxyde de manganèse, Rep. CEA R. 3284 (1967).

[82] YAMAGATA, N., IWASHIMA, K., Monitoring of sea water for important radioisotopes released by nuclear reactors, Nature (London) 200 (1963) 52.

[83] WATARI, K., TSUBOTA, H., KOYANOGI, T., IZAWA, M., Nippon Genshiryoku Gakkai Shi 8 (1966) 182-85.

[84] IKEDA, N., KIMURA, K., IZAWA, K., YASUI, T., Radioisotopes (Tokyo) 18 (1969) 8-11.

[85] HAMPSON, B.L., Analyst 88 (1963) 529-33.

[86] MITCHELL, N.T., Radioactivity in Surface and Coastal Waters of the British Isles 1970, Fisheries Radiobiological Laboratory, Lowestoft, Tech. Rep. FRL8 (1971) 32 pp.

[87] TSURUGA, H., Bull. Japan. Soc. Sci. Fish. 31 (1965) 651-58.

[88] CHAKRAVARTI, D., LEWIS, G.B., PALUMBO, R.F., SEYMOUR, A.H., Nature (London) 203 (1964) 571-73.

[89] SILKER, W.B., RIECK, H.G., Battelle Institute, Pacific Northwest Laboratory, Ann. Rep. 1968, BNWL-1051, Part 2 (1969) 18-20.

[90] DUTTON, J.W.R., "A review of methods for the determination of radioactive and stable silver in the marine environment", this report.

[91] ROBERTSON, D.E., Role of contamination in trace element analysis of sea water, Anal. Chem. 40 (1968) 1067-72.

[92] SCHUTZ, D.F., TUREKIAN, K.K., The investigation of the geographical and vertical distribution of several trace elements in sea water using neutron activation analysis, Geochim. Cosmochim. Acta 29 (1965) 259-313.

[93] ROBERTSON, D.E., The adsorption of trace elements in sea water on various container surfaces, Anal. Chim. Acta 42 (1968) 533-36.

[94] HARVEY, B.R., WOODS, S.M., Fisheries Radiobiological Laboratory, Lowestoft, Tech. Memo 3 (1971).

[95] GORSUCH, T.T., Analyst 84 (1959) 135-73.

[96] PIJCK, J., HOSTE, J., GILLIS, J., In Proc. Symp. Microchem., Birmingham Univ. 1958, Pergamon Press, Oxford, New York, London, Paris (1960) 48-58.

[97] GLEIT, C.E., HOLLAND, W.D., Anal. Chem. 34 (1962) 1454-57.

[98] HAMILTON, J.H., MINSKI, M.J., CLEWRY, J.J., Sci. Total Environ. 1 (1972) 1-14.

[99] REYNOLDS, E., Nature (London) 210 (1966) 615-16.

[100] FUKAI, R., HUYNH-NGOC, L., Variation of silver in the sea water around Monaco, J. Oceanogr. Soc. Japan 27 (1971) 91-100.

[101] BOWEN, H.J.M., Analyst 92 (1967) 124-31.

[102] DUTTON, J.W.R., "A review of methods for the determination of iodine-131 in the marine environment, this report.

[103] WORLD HEALTH ORGANIZATION, Method of Radiochemical Analysis, WHO, Geneva (1966) 94-103.

[104] PERLMAN, I., MORTON, M.E., CHAIKOFF, I.L., J. Biol. Chem. 139 (1941) 433.

[105] UNITED KINGDOM ATOMIC ENERGY AUTHORITY, P.G. Rep. 204 (1961) 12 pp.

[106] MORGAN, A., MITCHELL, G.R., Rep. AERE-M1004 (1962) 12 pp.

[107] KLEINBERG, J., COWAN, G.A., The Radiochemistry of Fluorine, Chlorine, Bromine and Iodine, Nat. Acad. Sci. - Nat. Res. Counc. Rep. NAS-NS 3005 (1960).

[108] GOLDBERG, E.D., "Minor elements in sea water", Chemical Oceanography (RILEY, J.P., SKIRROW, G., Eds) 1, Academic Press, London, New York (1965) 163-96.

[109] BHAT, I.S., KAMATH, P.R., GANGULY, A.K., Rep. BARC-644 (1972) 248.

[110] EICKE, H.F., "A method for rapid enrichment and determination of silver-110m and iodine-131 from seawater", this report.

[111] EDWARD, R.R., Science 137 (1962) 851-53.

[112] RUSSELL, J.L., HAHN, P.B., Radiological Health Data and Reports 12 (1971) 189-94.

[113] KOHMAN, T.P., EDWARD, R.R., Rep. NYD-3624-1 (1966) 43 pp.

SUPPORTING PAPERS

QUALITY CONTROL IN RADIOCHEMICAL ANALYSIS

J. H. HARLEY, H. L. VOLCHOK
Health and Safety Laboratory,
US Atomic Energy Commission,
New York, N.Y.

V. T. BOWEN
Woods Hole Oceanographic Institution,
Woods Hole, Mass.,
United States of America

Abstract

QUALITY CONTROL IN RADIOCHEMICAL ANALYSIS.
Whenever analytical data are to be used as a basis for a scientific study or even for monitoring purposes it is necessary to have information on the quality of the analyses. It is much better if this quality can be expressed in numerical form rather than subjectively and it is also better if the data are part of a continuing programme. All laboratories are involved in this question of analytical quality but the degree to which they develop the required information is highly variable. The present paper attempts to describe an ideal system and to present some data relating to analysis of seawater.

1. PROGRAM

It might be worthwhile to review here the overall requirements for quality control in the particular case of low level radiochemical analysis. It must be understood that a suitable program adds to the work required of the analyst and thus increases the cost. In return, the increased confidence that can be placed in the results far outweighs the economic disadvantage.

There are several basic factors which affect the quality of a radiochemical analysis.

(a) The use of proper standards for calibration.

(b) The use of proper counter efficiencies and backgrounds.

(c) The proper determination of radiochemical recovery.

(d) Correction of results for analytical blank.

(e) The continual checking of the performance of the overall system for accuracy and precision.

The required standards for low level radiochemical analysis have not always been available. National standardizing bodies are more concerned with highly exact calibration of high level instruments rather than furnishing solution standards for radiochemical analyses. Commercial sources have not always been reliable but our laboratory has had very satisfactory experience with the Amersham group in Great Britain and the CEA in France.

The IAEA has furnished a number of standard radionuclides and their quality
has been excellent. In the cases where none of these sources have the required standards it has been necessary to perform our own primary standard-
izations and to look to intercomparisons with other laboratories to confirm
the values.

The quality control of the final measurement process in radiochemical
analysis is relatively simple and has been described many times. The
most important features are maintenance of a quality control chart on
each individual counter and the use of counter efficiencies and backgrounds
averaged over an extended period. Additional counter data can be obtained
simply by use of a fixed stable counter standard which is not related to
the analysis in question but merely checks the reproducibility of counting.
An added control chart on such a standard can determine whether a shift
in calibration is due to the counter itself or to the technique of preparing the calibration standard.

The determination of radiochemical recovery has also been covered
many times in laboratory manuals. In the best cases isotopic tracers are
available for the radionuclide determined. Strontium-85, lead-212,
plutonium-236 and many others have been extremely helpful. In the case
of final determination of plutonium by alpha counting, the high quality
of the present alpha spectrometers allows recovery measurement with every
sample. Where suitable radioactive tracers are not available or the tracer
would interfere with the final counting the analyst must depend on carrier
recovery measurements. In any case, recovery should be tested with each
sample. It has been common practice for many routine analyses to measure
the recovery for a group of samples and to use the average recovery to
correct all future samples. This obviously has the defect that it will
not correct for changes in analytical technique or other errors.

The estimation of true blank to be subtracted from the value obtained
for a sample is frequently a very complex problem. The straightforward
measurement of a reagent blank may fail unless the proper elements are
present to generate the usual precipitates and go through the other radiochemical steps. Ideally the proper true blank would have all the constituents of the sample except for the one sought. This combination is
difficult to find and recourse has been made to the reuse of samples
carried through a separation procedure. The other value of blank determinations is the opportunity to detect contamination of new batches
of reagents. The requirements for blank material in this case are less
stringent since the blank value obtained is merely used for quality
control and is not subtracted from the sample activity.

The final check on the quality of analytical performance lies in
the test of the complete system. This means that each group of samples
submitted to the analyst should contain samples designed to test performance. These should include blanks and standard materials and they
should be submitted without identification. In this system, it is also
possible to use blind duplicates to test the precision of the process.
In the Health and Safety Laboratory we aim to include about 15% of quality
control samples in our laboratory operations. The same procedure is used
whether analyses are performed by our own chemists or under contract to
commercial laboratories. The percentage of quality control samples tends
to be larger with small batches of samples and perhaps somewhat smaller
with larger batches. This program gives a continuing record of the quality of analysis.

2.   OPERATION

The data from a quality control program are intended to provide a basis for action when the results show diminished quality. The required decisions are often difficult, and this may contribute to the unpopularity of quality control in some laboratories.

Problems with standards are only uncovered in the course of laboratory intercomparisons, such as those run by the IAEA. In most cases, it is possible to correct older data if the original standardization was incorrect. If the standard has shifted, the only proper procedure is to discard all data obtained in the questionable period. This requires an unusual degree of courage.

The control charts for counter efficiencies and backgrounds give immediate indications of any problems. Fortunately, out-of-control results do not require discarding data, but merely checking counter operation and possible maintenance service.

When radiochemical recovery values or quality control blanks give unsatisfactory results, while the counters are within control limits, the problem must be in the radiochemistry. Any decision on handling data from such a sample batch must depend on locating the cause of the poor results. The data from these measurements would lend themselves to quality control charting, but this has not been standard practice at HASL.

The greatest difficulties arise in the quality control samples for the overall system, standard samples, blanks and blind duplicates. In most series of low level radiochemical analyses, the quality control program has a serious defect. The time between submission of samples and receiving the results tends to be a matter of weeks and thus any problems detected cannot be corrected before additional samples are started. This raises the question of how to handle a batch of data when the quality control results are unsatisfactory. It is obviously statistically possible for a quality control sample to give a poor result while the rest of the analyses are correct. This means that running a single quality control sample with a batch is adequate only if you are willing to risk discarding the results of all analyses in the batch. If one out of two quality control results is unsatisfactory it is possible to accept the batch of data tentatively and see whether the following batch comes up to expected quality. These questions of data handling of course are entirely aside from using the results of quality control measurements to correct problems in the analysis. Incorrect standards, blanks or blind duplicates should all indicate that possible action is required to return the quality of analysis to the standards set by the laboratory.

3.   EXPERIENCE

The Health and Safety Laboratory (HASL) has supervised a program for Sr-90 analyses in seawater for a number of years. The seawater samples are collected primarily from the Atlantic Ocean under the scientific direction of Dr. Vaughan T. Bowen as part of his research activities at Woods Hole Oceanographic Institution (WHOI). HASL has the responsibility for obtaining qualified commercial contractors to carry out these analyses and to monitor the progress and quality of the results.

In general, this program has involved some 200 to 400 samples per year, this year we may do as many as 700. Typically, the sample volumes

are about 55 liters and the range of Sr-90 concentration is from zero to
0.5 picocuries per liter (pCi/l).

Choosing of the contractor, each year, is the initial step in the
quality control program. The selection is made after receipt of proposals
from qualified commercial analytical laboratories. Qualification is established by analysis of 3 to 5 samples of seawater, supplied by HASL.
These samples cover the range of concentrations expected in the program.
Contractor selection is made through evaluation of a number of criteria
such as: experience of both supervisory personnel and technicians, details of their radiochemical procedure, description (and if necessary an
inspection) of facilities and reasonableness of costs.

After a contractor is chosen samples are shipped from WHOI, generally
in batches of 100. Quality control, during the analytical phase of the
work, is maintained through legal constrictions written into the contract
and by special quality control samples submitted "blind" to the analysts.
Examples of the contractual controls are: requiring notice if key personnel (named in the contract) leave or are taken off this work, and, analyz-
ing the samples and reporting the data in numerical order. This last,
maintains the ratio of quality control to unknown samples in the analytical
sequence.

The special quality control samples have been designed to provide
specific information on: (a) accuracy, through analysis of standards or
knowns, (b) precision, through analysis of duplicates and repetition of
the standard, and (c) contamination, through analysis of blanks.

(a) <u>Accuracy</u>. In this program accuracy is determined through the
use of real seawater samples of known Sr-90 concentration, rather than by
analysis of prepared synthetic standards. The reasoning behind this decision involves the difficulty in preparing a seawater standard in sufficient quantity to be useful in a program of this size, plus the question
of whether a valid matrix could be synthesized. Our procedure is to
obtain, from a convenient location, a large volume of real seawater and
establish the Sr-90 concentration by analysis at a number of experienced
laboratories. One unidentified 55 liter drum of this water is then submitted to the contractor with each group of 10 samples. In the early
years, we obtained our standard from a deep well used to supply water to
the aquarium at Coney Island, New York. Later, for reasons not clear as
yet, the Sr-90 concentration in that water rose substantially above that
found in the mid-latitude Atlantic and this source was therefore abandoned.
For the past several years we have used open ocean surface water, taken
not too distant from WHOI. For the most part, since inception of this work
the analysts have maintained an accuracy of within ± 10% in measurement
of these standards, at concentrations of about 0.2 to 0.5 pCi/l.

(b) <u>Precision</u>. Two methods of measuring the precision of these seawater analyses are employed. First, since in each year, several hundred
samples are analyzed, the values of the standard (one out of every 10
samples) provides a very good indication of precision. Second, in each
batch of 10 samples there is a pair of duplicates. These are either aliquots from a large sample or two samples obtained side by side at sea.

It should be pointed out that these two precision indicators are not
identically matched in that the standard always represents surface water
concentration (or higher) while the duplicate pairs vary in concentration
from that level to almost background. Table 1 summarizes the data on
precision from four different analytical contractors, since 1968. It

seems quite clear that the precision of Sr-90 analysis in seawater is
quite acceptable when compared to similar data from other programs concerned with smaller samples and less complicated matrices than seawater.
Further the difference between the precision as determined from standards
and that from the duplicates is probably attributable to the often lower
concentrations found in the latter, as mentioned earlier. It must be
emphasized that this excellent record outlined in Table 1, belongs to
only a very few laboratories, carefully selected by the procedure described.

(c) Contamination. Overall contamination of samples during the
analytical phase of the program is observed by use of blanks. A true
ideal blank, as defined previously, is rarely obtainable in environmental
or geophysical research, and seawater analysis is no exception. There
simply is no source of real seawater which is free of Sr-90. In the
beginning of the program we attempted to manufacture a blank from a com-
mercially obtained sea salt mixture. This was unsuccessful because the
salts did not contain many of the trace elements found in seawater
(notably strontium) and the processing of the mixture was very difficult.
We currently use a seawater blank consisting of saline water from a Gulf
Oil Corporation oil field near Roswell, New Mexico. The composition of
this water is such that with only simple dilution it is quite acceptable
for the purpose. Of course one of the criteria for choosing this
particular oil field as the source was the knowledge that the water could
not have been exposed to possible Sr-90 contamination. Because of the
difficulty in obtaining these samples, we have averaged only about one
blank per 30 samples. Additionally, however, we require that the analyst
run one reagent blank for each 25 samples. Blank values reported by our
contractors in almost all cases have been less than 0.005 pCi/1.

Our experience in this program has given us a great deal of confidence in the power of quality control to help maintain a high standard of
output. We would emphasize here that, in this particular type of work,

Table I

Precision of Sr-90 Analysis in Seawater
(number of cases in parentheses)

Laboratory	Period	Standards (a)	Duplicates (b)
A	1968 - 1969	3.5% (10)	4.9% (10)
B	1969 - 1970	3.3% (11)	5.0% (10)
C	1970 - 1971	7.6% (16)	11.7% (13)
D	1971 - 1972	4.0% (17)	7.8% (14)

(a) Standard deviation of all analyses of the standard during
the contract period.

(b) Average deviation between duplicate pairs during the
contract period.

the most critical element in the quality control system is the attention
of the supervisory personnel in the laboratory. Our only serious difficulties have come about when, due to personnel changes, a new supervisor
without adequate experience was used. Hence, we now insist upon strict
control of this aspect, and will not allow analysis of our sample until
satisfied that the right person is in charge.

    In closing, we believe strongly in quality control as an essential
part of quantitative radiochemistry. All laboratories occasionally drift
off calibration, all chemists occasionally switch samples or miscalculate
results. A good quality control program serves as a continuous surveillance system to alert us to possible mistakes.

DOUBLE-TRACER STUDIES TO OPTIMIZE CONDITIONS FOR THE RADIOCHEMICAL SEPARATION OF PLUTONIUM FROM LARGE SEAWATER SAMPLES

H.D. LIVINGSTON, D.R. MANN, V.T. BOWEN
Woods Hole Oceanographic Institution,
Woods Hole, Mass.,
United States of America

Abstract

DOUBLE-TRACER STUDIES TO OPTIMIZE CONDITIONS FOR THE RADIOCHEMICAL SEPARATION OF PLUTONIUM FROM LARGE SEAWATER SAMPLES.

The results of experiments using double-tracer techniques with ^{236}Pu and ^{242}Pu to optimize conditions in the radiochemical separation of plutonium from large volume seawater samples are presented. These show that the major areas of loss are the iron hydroxide precipitation step, the adsorption of plutonium on the anion exchange resin and the final electrodeposition procedure. Data are presented that show that the use of increased weight of iron 0.5 - 1 mg per sample greatly improves the recovery. The use of hydrogen peroxide to ensure oxidation of the plutonium to the 4+ state is also recommended to improve retention on the anion resin. It is well known that the 3+ state of Pu is little adsorbed. Electrodeposition can also be improved using ammonium sulphate and longer plating times at smaller current densities.

The determination of the concentrations of plutonium introduced to the world's oceans as fall-out from nuclear testing poses an analytical task of considerable magnitude. The amounts found in typical contemporary seawater lie for most part in the range 0-1 dpm/100 kg for $^{239/240}$Pu [1]. Despite the sensitivity now available for alpha-particle measurement through the use of surface barrier detectors and pulse-height analysis, useful analyses still require the separation of radiochemically pure plutonium in high yield from quite large (e.g. 50 kg) seawater samples.

The radiochemical separation of plutonium from seawater is basically a two-step process:
(1) Concentration of plutonium from seawater
(2) Purification of the concentrated plutonium for α-spectrometry.

Several procedures have been described that use a variety of different approaches to this two-step procedure [2-8]. Since observed recoveries of plutonium (measured using plutonium tracer as yield monitor) can show only the product of all the various losses that occur at each step in a multi-step procedure, it is difficult to identify the key factors that affect the quantitative recovery of plutonium at individual steps. As a result, misleading or erroneous conclusions are easily made about the effect of these factors. When plutonium separation from seawater is not the sole analytical objective but is only a part of a more general, sequential radiochemical analysis, the difficulties are compounded.

This indeed is the situation that existed in our laboratory. Two variations have been described on a generalized plutonium separation scheme incorporated into a pre-existing scheme for analysis of ^{90}Sr, ^{137}Cs and other fission product nuclides in seawater [2, 3]. Unsatisfactory or unpredictable radiochemical yields of plutonium forced us to re-evaluate our analytical procedures to establish the cause or causes of unacceptable radiochemical losses. The technique of addition of a second plutonium tracer (using an isotope neither sought nor added as the first tracer) at a selected point in the analytical scheme enabled us to discover some factors that were adversely affecting our yields. Plutonium-242 is the isotope that replaced the ^{236}Pu described previously as the plutonium yield monitor. The former is a more satisfactory tracer because:

(1) Its longer half-life removes the need for continual decay correction, as required for ^{236}Pu;

(2) Its decay products are sufficiently long-lived to be still essentially insignificant even after several years decay of their parent, in contrast to those of ^{236}Pu; and

(3) Its alpha-particle energies are below those of $^{239/240}$Pu and ^{238}Pu (the isotopes being sought) and consequently interference caused by energy degradation tailing is avoided.

In all of the analyses described here ^{242}Pu (about 2 disintegrations per minute, carefully calibrated) was added to seawater samples (50 to 60 litres), along with various carriers, and was used to determine the overall recovery of plutonium separated from the sample. Then recovery of ^{242}Pu at any chosen point in the procedure was measured by the addition of a similar amount of ^{236}Pu at that point and the percentage of ^{236}Pu finally recovered was used to calculate the ^{242}Pu recovery at the time of ^{236}Pu addition. An example of how this was used can best serve as illustration of the technique.

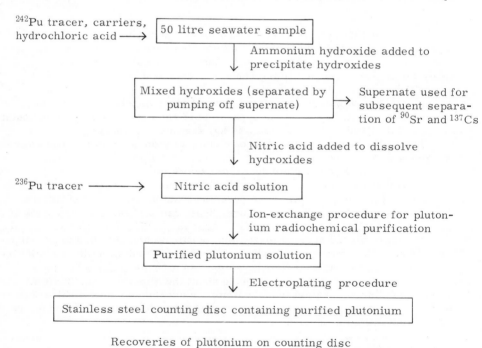

^{242}Pu tracer, carriers, hydrochloric acid ⟶ | 50 litre seawater sample |

Ammonium hydroxide added to precipitate hydroxides

| Mixed hydroxides (separated by pumping off supernate) | ⟶ Supernate used for subsequent separation of ^{90}Sr and ^{137}Cs |

Nitric acid added to dissolve hydroxides

^{236}Pu tracer ⟶ | Nitric acid solution |

Ion-exchange procedure for plutonium radiochemical purification

| Purified plutonium solution |

Electroplating procedure

| Stainless steel counting disc containing purified plutonium |

Recoveries of plutonium on counting disc

^{242}Pu	^{236}Pu
50%	75%

Therefore the recovery of ^{242}Pu at the time of ^{236}Pu addition was $50/0.75 = 66.7\%$. From this it is evident that the initial concentration of plutonium from the 50 litres of seawater was only 66.7% effective, i.e. $100-66.7 = 33.3\%$ of the activity remained in the supernate.

The ^{236}Pu second tracer was also used to examine specific locations where ^{242}Pu losses were suspected. For example, in the ion-exchange procedure used in plutonium purification the initial step is the adsorption of newly oxidized Pu^{IV} as a nitrate complex on anion exchange resin. Any plutonium remaining in the Pu^{III} oxidation state is scarcely adsorbed by the resin and this effect can cause a loss of plutonium in the overall scheme. The extent to which this mechanism was actually contributing to overall losses was determined by addition of ^{236}Pu to the column eluate, re-oxidation of plutonium to Pu^{IV} followed by the same purification and plating procedure used for a sample. The amount of ^{242}Pu present in the original column eluate could be determined by correction for ^{236}Pu recovery of any ^{242}Pu measured on the counting plate.

In this way the extent of plutonium loss at the various stages in our procedures [2, 3] was examined in reverse order. Initially the electroplating method was evaluated, secondly the ion-exchange procedure and finally the concentration of plutonium from the original sample of seawater.

ELECTROPLATING

The electroplating method we have described previously [2] uses an ammonium chloride medium. The plating recoveries of six samples by this method (15 min, 2 amp) were $86 \pm 5\%$. Studies of this method indicated that better recoveries resulted from longer plating times and lower current. In addition, use of an ammonium sulphate medium produced high-quality plates and higher recovery. Eight samples plated for 2 hours at 1 ampère, the conditions that we use at present (based on studies by Talvitie [9]), gave recoveries of $97 \pm 4\%$.

ION EXCHANGE

The procedure described by Wong [2] includes an adsorption step on an anion exchange resin of Pu^{IV} from 8M nitric acid. Our suspicion that some Pu losses resulted from incomplete adsorption at this stage were confirmed when re-analysis of the adsorption effluent of some samples showed that variable amounts of Pu could be found there. The significance of this observation was strengthened by some experiments examining this step in closer detail. In four experiments the adsorption, under various conditions, of a known amount of ^{242}Pu tracer on the resin was measured by exhaustive elution of the adsorbed ^{242}Pu, addition of ^{236}Pu to monitor the electroplating yield, plating and α-spectrometry. The conditions used included the presence or absence of Fe and the use or non-use of hydrogen peroxide prior to the sodium nitrite oxidation of Pu, as previously described. The amount of Pu adsorbed was in the range 42-59% in the four experiments — much less than expected. In two of the experiments the adsorption effluent was electroplated and counted to confirm that the missing Pu had passed through the anion column.

TABLE I. FACTORS AFFECTING COPRECIPITATION OF PLUTONIUM FROM SEAWATER WITH HYDROXIDES

Factor being evaluated	Fe oxidation state	Amount of carrier (g)	Order of Fe precipitation	No. of samples	Plutonium recovery (%)		
					Mean ± 1σ	Median	Range
Oxidation state of iron	II	0.1	1	6	53 ± 3	53	50-57
	III	0.1	1	7	60 ± 9	58	47-74
	II	0.1	2	8	49 ± 11	50	32-60
Amount of iron	II, III	0.1	1	13[a]	57 ± 8	55	48-73
	III	0.3	1	1	73	-	-
	III	0.4	1	1	77	-	-
	III	0.6	1	1	61	-	-
	III	0.8	1	1	79	-	-
	III	1.0	1	4	78	83	56-90
	III	3.0	1	1	93	-	-
	III	9.0	1	1	100	-	-
	II	0.1	2	8[a]	49 ± 11	50	32-60
	III	0.5, 1.0	2	7	68 ± 6	68	68-70
Order of precipitation	II, III	0.1	1	13[a]	57 ± 8	55	48-73
	II	0.1	2	8	49 ± 11	50	32-60
	III	0.5-1.0	1	6[a]	75 ± 13	81	56-90
	III	0.5-1.0	2	7[a]	68 ± 6	68	60-78

a Repetition of data already tabulated.

Talvitie [10] reported that Pu could be oxidized to Pu^{IV} with hydrogen peroxide and adsorbed (with high efficiency) on anion exchange resin from strong nitric or hydrochloric acid media. We were able to confirm this report in a series of simulated samples. Using similar oxidation conditions [10] (1 drop 30% hydrogen peroxide per 10 ml of acid), recoveries in the range 83-106% were obtained. We also found, like Talvitie, that slightly higher adsorption could be obtained from hydrochloric acid than from nitric acid. Because of the amount of iron we use to coprecipitate Pu from 50 litres of seawater, we found it more convenient to continue using nitric acid as the acid from which Pu is adsorbed on the resin.

Our experience since changing our Pu oxidation and resin adsorption procedures to those described by Talvitie has been that with real 50-litre seawater samples spiked with a second Pu tracer prior to ion-exchange separation it is possible with single column separation and electroplating to recover the second Pu tracer with yields around 75%.

CONCENTRATION OF PLUTONIUM FROM SEAWATER

Some factors were studied that affect the coprecipitation of plutonium with iron hydroxides precipitated from large seawater samples. These factors included the amount and oxidation state of the iron and the effect of making the hydroxide precipitation before or after Cs removal with ammonium molybdophosphate (AMP). Wong [2] concluded that plutonium was much more efficiently coprecipitated from large volumes of seawater when the iron was precipitated in the Fe^{II} rather than the Fe^{III} state. He also stated that the AMP separation of caesium did not carry detectable amounts of plutonium and consequently this separation could be made before the hydroxide coprecipitation of plutonium without significant loss of the latter. The double-tracer experiments described here contradict these two observations.

FACTORS AFFECTING COPRECIPITATION OF PLUTONIUM FROM SEAWATER WITH HYDROXIDES

Oxidation state of iron

In the series of samples studied to evaluate the effect of this variable hydroxylamine hydrochloride (14 g) was added, as a reducing agent for iron, to a number of samples of seawater before making the hydroxide precipitation. In all cases the amount of iron present was 0.1 g. The hydroxides formed in these samples were blue-green in colour, indicating that the iron was predominantly in the Fe^{II} state as it precipitated. This was in contrast to the red-brown Fe^{III} hydroxide formed when the reducing agent was not used. This series of samples included some in which the hydroxide fraction was the first separated from the seawater, and some where it followed Cs removal by AMP. The results from these analyses are shown in Table I. The plutonium coprecipitated on the hydroxide precipitate was measured as described above from double-tracer experiments. The data show that this recovery figure was not affected by the oxidation state of the iron during precipitation. If anything, the average recovery was slightly

higher when the iron was precipitated in the Fe^{III} state. The slightly lower
recoveries found in those samples where the Fe^{II} hydroxide was formed
after Cs removal are consistent with the losses attributable to the AMP
described below.

Amount of iron

The quantity of iron used to coprecipitate plutonium from 50 litre
seawater samples was varied in a series of samples. Hydroxide precipitation was the first separation step in 23 of the 30 samples included in this
series. The data in Table I show that the recovery of plutonium from the
seawater increased with the amount of iron carrier used. On the basis of
these data it seemed worthwhile to use as much iron as possible. We settled
on 0.5 g as a compromise between the increased plutonium recoverable and
the handling problems associated with the voluminous hydroxides obtained
when 1 g or more were used. A similar relationship between plutonium
recovered and the amount of iron used was found when the hydroxide
precipitation was made after Cs removal. From eight samples processed
using 0.1 g of iron 49 ± 11% of the plutonium was coprecipitated with the
hydroxides. When 0.5 or 1g of iron was used in a further seven samples,
the average plutonium coprecipitated increased to 68 ± 6%. Although the
oxidation state of the iron differed in these two sets, they can still stand
comparison since this factor was shown above not to affect plutonium recovery
Again the lower recoveries found when hydroxide precipitation followed Cs
removal are consistent with the amounts of plutonium that can be lost
during Cs removal.

Order of hydroxide precipitation

As indicated above, the separation of Cs with AMP from seawater
samples was found to remove significant quantities of plutonium. This was
shown in two ways.

(1) Plutonium found with Cs fraction

The Cs-AMP slurry removed from a seawater sample is dissolved in
10$\underline{M}$ sodium hydroxide solution [2]. The residue remaining after solution of
the Cs-AMP slurry is usually discarded. In a series of samples these
residues were leached with 16$\underline{M}$ nitric acid and a known quantity of ^{236}Pu
tracer added. After radiochemical purification of this fraction and
α-spectrometric analysis the quantity of the original ^{242}Pu tracer present
in the Cs-AMP slurry was calculated. In nine samples treated in this
fashion the average amount of ^{242}Pu found in the slurry was 8 ± 6% of that
initially added to the seawater.

(2) Plutonium recoveries with hydroxides

The recoveries of plutonium at the hydroxide precipitation stage were
compared in two series of samples — in one hydroxide precipitation was
before, in the other after Cs separation on AMP. When 0.1 g of iron was
used, thirteen samples averaged 57 ± 8% when hydroxides were separated
first (Table I), whereas eight samples where Cs was first separated averaged

TABLE II. EFFECT OF PRIOR HYDROXIDE PRECIPITATION ON CHEMICAL RECOVERY OF Cs

Operator	Order of iron precipitation	No. of samples	Cs recovery (%)	
			Mean ± 1σ	Range
1	First step	24	49 ± 7	33-60
1	Second step	16	74 ± 16	40-99
2	First step	27	46 ± 9	23-61
2	Second step	52	81 ± 8	56-90

49 ± 11%. The recoveries when 0.5 - 1 g iron were used were 75 ± 13% (six samples) when hydroxides were precipitated first, and 68 ± 6% (seven samples) when Cs was removed first. Clearly losses of plutonium are liable to occur if Cs is removed by AMP before hydroxide precipitation.

Although higher plutonium recoveries can be achieved by separating plutonium with hydroxides prior to Cs separation, this separation order severely reduced Cs recoveries — presumably Cs was lost by coprecipitation with the hydroxides. These data are shown in Table II and are from samples analysed by two different operators. The data from both analysts show that the average Cs recovery was reduced by almost a factor of 2 when Cs-AMP separation followed hydroxide precipitation. Since the reduction of Cs recoveries was much more severe following hydroxide precipitation than was the reduction of Pu recoveries following Cs-AMP separation, the preferred order of separation when both Cs and Pu are being determined is (1) Cs, (2) Pu.

The experience gleaned from these double-tracer experiments has permitted improvements in our ability to recover plutonium from 50 litre seawater samples for α-spectrometric analysis of $^{239/240}$Pu and, in some cases, ^{238}Pu. Plutonium recoveries in excess of 50% have been recently routine, rather than exceptions as was the case prior to this study of the sources of losses during concentration and purification.

SUMMARY

By using two tracers of a single element as yield-monitors, adding one at late steps in a sequential procedure, it is possible straightforwardly to examine the conditions affecting efficiency at each step. This has been illustrated by our experiments in optimizing our procedure for plutonium in seawater, using ^{242}Pu and ^{236}Pu for double-tracer studies. The results shown also serve to justify the specific details of our present procedure, as set out elsewhere [11].

REFERENCES

[1] NOSHKIN, V.E., Ecological aspects of plutonium in aquatic environments, Health Phys. 22 (1972) 537-49.
[2] WONG, K.M., Radiochemical determination of plutonium in sea-water, sediments and organisms, Anal. Chim. Acta 56 (1971) 355-64.

[3] WONG, K.M., NOSHKIN, V.E., BOWEN, V.T., "Radiochemical methods for strontium-90 and other fallout nuclides in sea-water", Reference Methods for Marine Radioactivity Studies, Technical Reports Series No. 118, IAEA, Vienna (1970) 119-127.

[4] HOLLSTEIN, U., HOOGMA, H.M., KOOI, J., An improved method for the determination of trace quantities of plutonium in aqueous media — II. Application to sea-water and urine, Health Phys. 8 (1962) 49-57.

[5] PILLAI, K.C., SMITH, R.C., FOLSOM, T.R., Plutonium in the marine environment, Nature (London) 203 (1964) 568-71.

[6] PILLAI, K.C., SMITH, R.C., FOLSOM, T.R., The Separation and Determination of Plutonium in Sea-water and Marine   Organisms, Environmental Marine Radioactivity Series, Rep. IMR-TR-922-66-B (1966).

[7] PILLAI, K.C., "Determination of plutonium in the marine environment", these Proceedings.

[8] AARKROG, A., "Radiochemical determination of plutonium in marine samples by ion exchange and solvent extraction", these Proceedings.

[9] TALVITIE, N.A., Electrodeposition of actinides for alpha spectrometric determination, Anal. Chem. 44 (1972) 280-89.

[10] TALVITIE, N.A., Radiochemical determination of plutonium in environmental and biological samples by ion exchange, Anal. Chem. 43 (1971) 1827-30.

[11] BOWEN, V.T., LIVINGSTON, H.D., NOSHKIN, V.E., "Methods of analysis of plutonium in marine environmental samples, with some remarks concerning americium, neptunium and curium", presented at Reference Methods for Marine Radioactivity Studies — Determination of Transuranic Elements, Radioruthenium and other Radionuclides in Marine Environmental Samples, Working Group Meeting, 30 Oct. - 3 Nov. 1972, Vienna.

A PROCEDURE FOR ANALYSIS OF AMERICIUM
IN MARINE ENVIRONMENTAL SAMPLES

R. BOJANOWSKI*, H. D. LIVINGSTON, D. L. SCHNEIDER, D. R. MANN
Woods Hole Oceanographic Institution,
Woods Hole, Mass.,
United States of America

Abstract

A PROCEDURE FOR ANALYSIS OF AMERICIUM IN MARINE ENVIRONMENTAL SAMPLES.
 A method is described for the measurement of americium in marine environmental samples. Some data
are presented to show the method is capable of acceptable accuracy and precision. An extensive series of
notes discusses the rationale for various steps in the procedure.

The measurement of significant amounts of various plutonium isotopes
in the marine environment [1, 2, 3, 4] has demonstrated that other transuranics should be present in detectable amounts. Americium-241 ($T\frac{1}{2}$ 458
years) is produced by β-decay of ^{241}Pu ($T\frac{1}{2}$ 13.2 years) and consequently
not only is introduced to the environment directly in fall-out but in-
directly by 'in situ' generation from ^{241}Pu over a time interval approaching a century. The method described here can be used to determine, and to
compare with plutonium, the marine levels and biogeochemistry of ^{241}Am.

As in plutonium analytical methods, an α-emitting radiotracer, ^{243}Am,
is used to measure the radiochemical yield of ^{241}Am separated from a
sample. Because the α-energies of ^{243}Am (5.28 MeV, 87%; 5.23 MeV, 11.5%)
and ^{241}Am (5.49 MeV, 85%; 5.44 MeV, 13%) are only just resolved by high-
performance surface barrier detectors, it is important that the amount of
tracer activity added to a sample be greater than that of the ^{241}Am present - to reduce inaccuracy caused by 'tailing' of the higher energy ^{241}Am
peak into or even over the ^{243}Am tracer peak.

Since the chemistry of americium is so similar to the lanthanides, a
lanthanide carrier is used in preliminary concentration of americium from
sample matrix elements. Various combinations of hydroxide and oxalate
precipitations are used towards this end. After the concentration steps,
the lanthanide/americium fraction is radiochemically purified by ion exchange from α-emitting isotopes of the natural radioactive series.

Americium must be completely purified from the lanthanide carrier prior
to electroplating americium for α-spectrometry as the lanthanide also
plates and interferes with the measurement. An anion exchange method of
purification is used to make this separation. The separation is based on
the difference in adsorption from thiocyanate media between americium and
the lanthanides. Americium is quantitatively adsorbed on an anion resin
column from 2M thiocyanate solution while the lanthanides pass through with
little adsorption. After the lanthanides have been completely removed,
americium is stripped from the resin with 4M hydrochloric acid. This fraction is subjected to an additional thiocyanate column treatment to remove

---

* Polish Academy of Sciences, Marine Station SOPOT-MOLO, Poland.

the last traces of the lanthanide carrier. The americium is pure enough
at this stage to be electroplated after destruction of residual thiocyanate
eluted with the americium. The activity of 241Am and of the 243Am yield-
monitor are then measured using a high resolution surface barrier detector
as previously described for plutonium [5].

The subsections of the procedure, as described below, are provided
with notes, that clarify the reasoning behind particular steps, or that
warn against special precautions required.

PROCEDURE

1. SAMPLE PREPARATION.

A) Sea-water: Follow the procedure described in "Reference Methods for
Marine Radioactivity Studies", [6] from 1-1 through 2-18. This describes
the separation of several radionuclides from 55 liter seawater samples by
coprecipitation with iron hydroxide. Americium radionuclides are concentrated from seawater by this procedure. An aliquot of standardized tracer
for Am (243Am, approximately 1-2 disintegrations per minute) is included in
Step 1-2. (See Note 1-a).

B) Biological and Sediment Samples: Preliminary treatment of biological
and sediment samples is made essentially as described in the procedure for
plutonium analysis [5], except that 242Pu has replaced 236Pu as the Pu
tracer. In this procedure appropriate tracers and carriers (now including
1-2 dpm of 243Am (standardized) and 50 mg Nd) are added to either the HNO_3/HCl
leach of sediment, or to ashed biological material. Plutonium is oxidized
to Pu (IV) and removed by adsorption from nitric acid solution on an anion
exchange resin. Americium is concentrated from solution after Pu removal
by one of the following procedures -- depending on the nature of the sample.

1. Hydroxide Precipitation: When the sample is calcium rich, but low in
phosphate.
2. Oxalate Precipitation: For low calcium samples.
3. Oxalate/Hydroxide Precipitation: For samples rich in calcium phosphate.

1. Hydroxide Precipitation

1-B1-1. Reduce by evaporation the volume of the 8M HNO_3 solution after Pu
removal.
1-B1-2. When salts begin to form, dilute the sample with water to 2-3 litres.
1-B1-3. Add 14M NH_4OH while stirring until the solution is strongly basic
(pH>10). Cover and leave the hydroxides to settle. (See Note 1-b).
1-B1-4. Remove the supernate, transfer the hydroxide slurry to a 250 ml
centrifuge bottle, centrifuge and discard the supernate. Wash the precipitate with dilute NH_4OH, centrifuge and discard the supernate.
1-B1-5. Dissolve the hydroxide precipitate in the minimal amount of 12M
HCl, dilute with water and centrifuge if necessary (Note 1-c) Decant the
supernate into a 1000 ml beaker, wash any precipitate with 0.1 M HCl, centrifuge, and combine wash solution with original supernate.
1-B1-6. Add 12M HCl to the solution to make it 3-4M, cover and boil for 5
min. Dilute with water to about 800 ml and heat to near boiling.(Note 1-d).
1-B1-7. Precipitate the hydroxides at pH >9 using 14M NH_4OH. Leave the
precipitate until it settles.
1-B1-8. Repeat Steps 1-B1-4 and 1-B1-5. (Note 1-e).
1-B1-9. Dilute with water and adjust pH to 1-2 using 14M NH_4OH. Add
an equal volume of saturated oxalic acid solution, mix well and let stand
for at least 1 hour (stirring occasionally), but preferably overnight.

1-B1-10. Decant the supernate, wash the precipitate with 0.5% oxalic acid, centrifuge and discard the wash solution.

1-B1-11. Dissolve the oxalates in 2ml 8M HNO_3, warming if necessary, dilute with water to about 30 ml, adjust pH to 1-2 with 14M NH_4OH and reprecipitate the rare earth oxalates by addition of 50 ml saturated oxalic acid (Note 1-f). After standing for 0.5 hour, centrifuge, wash oxalates with 0.5% oxalic acid and discard the supernate.

2. Oxalate Precipitation

1-B2-1. As for 1-B1-1 and 1-B1-2.

1-B2-2. Adjust the pH of the solution to 1-2 using 14M NH_4OH. Add solid oxalic acid to make the solution 5% with respect to oxalic acid. Stir to dissolve the oxalic acid, warming if necessary. If at this stage the precipitate does not form, add 14M NH_4OH dropwise until a definite turbidity appears. Let the precipitate form and settle overnight. (Note 1-g).

1-B-2-3. Discard the supernate and wash the precipitate with 0.5% oxalic acid. If transparent crystals of ammonium oxalate are present in the precipitate they should be dissolved in 0.5% oxalic acid (warm if necessary). Discard the wash solutions (Note 1-h).

1-B2-4. Transfer the precipitate to a suitable beaker and after heating, dissolve it in as little 8M HNO_3 as possible.

1-B2-5. Dilute the solution with about 10 times its volume of water and adjust pH to 1-2 with dilute NH_4OH. Add an equal amount of saturated oxalic acid, let stand for at least 0.5 hour, centrifuge, and wash the precipitate with 0.5% oxalic acid, discarding the supernate and washes. (Note 1-i).

3. Oxalate/Hydroxide Precipitation

1-B3-1. As for 1-B1-1 and 1-B1-2.

1-B3-2. Adjust pH of the solution to 1-2 using 14 M NH_4OH. Add solid oxalic acid to make the solution 5% with respect to oxalic acid. Stir to dissolve oxalic acid, warming if necessary. Add 14M NH_4OH to the solution until clear signs of precipitation occur. Let the precipitate form and settle overnight.

1-B3-3. Wash the precipitate with 0.5% oxalic acid, filter and ignite the mixed oxalate. Heat the crucible strongly after all carbon is volatilized to ensure proper conversion of oxalates to oxides.

1-B3-4. Dissolve the mixed oxides in minimal amount of 3M hydrochloric acid and transfer the solution to a centrifuge tube. Centrifuge and discard any insoluble material.

1-B3-5. Add 14M NH_4OH until hydroxides precipitate. Centrifuge and discard the supernate. Wash the hydroxide precipitate, centrifuge and discard wash solution.

1-B3-6. Dissolve the hydroxide in 12M hydrochloric acid, adjust pH to 1-2 and precipitate oxalates by addition of an equal volume of saturated oxalic acid solution.

1-B3-7. Continue as in 1-B1-10 and 1-B1-11.

NOTES

1-a. If analysis of rare earth nuclides is not required, 50 mg of Nd may be added instead of the rare earth carriers listed in 1-2 [6]. Nd is a good carrier for Am; it can be obtained in sufficient purity and, being colored, can be followed visually during precipitations, etc. Improvements over the procedure described in Reference Methods for Marine Radioactivity Studies [6] include:
   1) Use of ^{242}Pu as a Pu tracer instead of ^{236}Pu.
   2) Discontinuation of the use of a reducing agent prior to iron hydroxide separation [7].

1-b. The hydroxide precipitation should be made if possible from hot solution, to make the hydroxides settle better, allowing removal of the supernate by aspiration.

1-c. The precipitate may consist of easily hydrolyzable elements e.g. Ti, Si, Zr and any Th or Pu incompletely removed by the earlier anion exchange procedure.

1-d. This procedure should bring any colloidal hydroxides into solution.

1-e. It is important to minimize the acid used in dissolving the hydroxides because excessive amounts of NH_4^+ ions (from NH_4OH used in subsequent pH adjustment) interfere with oxalate precipitations by forming sparingly soluble ammonium oxalate.

1-f. This oxalate precipitation is a purification of the rare earth oxalates from coprecipitated calcium. Calcium oxalate does not convert to a hydroxide in the subsequent purification step.

1-g. The amount of calcium precipitated with the rare earth oxalates increases with increasing pH. But as more calcium precipitates, the rare earths precipitate more efficiently. The conditions of precipitation should be judged depending on the nature of an individual sample. However, an excess of NH_4^+ ions should be avoided and the pH of the solution should not be greater than pH 2. If a precipitate does not form even at pH 2, it would be better to add additional rare earth carrier to initiate precipitation rather than raise the pH above pH 2. Frequently, leaving the solution overnight helps precipitation.

1-h. Dilute oxalic or hydrochloric acid is a better choice of wash solution than water because the solubility of ammonium hydrogen oxalate is much higher than the neutral salt.

1-i. Upon addition of water the rare earth oxalates precipitate again. Nevertheless the pH adjustment and oxalic acid addition should still be made.

2. PURIFICATION PROCEDURE.

The rare earth oxalates with coprecipitated americium are further purified as follows:

2-1. Metathese the oxalates to hydroxides by addition of 5 ml of 10M NaOH. Break up the lumps so as to obtain a homogeneous slurry. Let stand for 5 minutes, mixing occasionally, dilute with water and centrifuge.

2-2. Discard the supernate, add a few drops of 10M NaOH, mix well, dilute with water and after 5 minutes, centrifuge. Remove and discard the supernate.

2-3. Dissolve the hydroxides with 10 ml 8M HNO_3, add 0.1 g $NaNO_2$ and let stand for 5-10 minutes.

2-4. Transfer the solution to a previously conditioned anion exchange column (Note 2-a).

2-5. Allow the solution to pass through the column at a flow-rate of 2-3 ml/min. and collect column eluate plus washings (using 50 ml 8M HNO_3).

2-6. Evaporate this fraction to dryness.

2-7. Dissolve the residue in 10 ml 8M HNO_3, add 0.1 g $NaNO_2$ and repeat steps 2-4 and 2-5.

2-8. Evaporate solution almost to dryness, transfer to a 20 ml beaker and continue to evaporate to complete dryness.

2-9. Dissolve residue in 2 ml 12M HCl and evaporate to dryness. Repeat this step.

2-10. Wash beaker well with deionized water (2 ml) and evaporate to dryness. Repeat this step.

2-11. Dissolve residue in 1 ml of 1.5 M HCl and transfer to a previously prepared anion exchange column (Note 2b). Rinse beaker twice with 1 ml of 1.5 M HCl and add to column.

2-12. Pass each fraction through the column and use 3 x 1 ml of 1.5 HCl for washing the column. Collect eluate and washings and evaporate to dryness.

NOTES

<u>2-a</u>. Dimensions of column: 1 cm x 20 cm. Resin: AG 21 K (BioRad Labora-
tories, Richland, California). Mesh size 50-100. Conditioned by passing
50 ml of 8M HNO_3 containing 0.1 g $NaNO_2$/100 ml at a flow-rate of 2-3 ml/min.
        This column is used to purify the Nd/Am fraction from residual traces
of Pu or Th.
<u>2-b</u>. Dimensions of column: 0.25 cm x 7 cm, (A disposable transfer pipette
can be used) Resin: AG 1 x 8, (BioRad Laboratories, Richland, California).
Mesh size 100-200. Conditioned by passing 5 ml of 1.5 M hydrochloric acid
at a flow-rate of 0.5 ml/min.
        This column adsorbs Pb, Bi and Po isotopes originating from the natu-
ral radioactive series and which may not be completely separated until this
stage. The hydrochloric acid concentration is critical for Pb adsorption
[8].

3. AMERICIUM/RARE EARTH SEPARATION.

        This is required because any rare earths present when americium is
being electroplated also deposit and interfere with the α-spectrometry.
<u>3-1</u>. Add 2 ml warm 6M NH_4SCN, mix, then neutralize excess acid <u>carefully</u>
by dropwise addition of 1M NH_4OH until the pink color (due to traces of iron)
just disappears. Restore the color by dropwise addition of 0.2M HCl. At
this stage the solution should be slightly acidic (pH >2), should not contain flakes of hydroxide and should be slightly pink (Fe^{III} thiocyanate).
The total volume should not exceed 5 ml.
<u>3-2</u>. Transfer the solution to the top of an anion exchange column (Note 3-a).
(Take care not to spread the sample over the walls of the column. Apply the
solution using a transfer pipet in small portions, letting each portion soak
into the resin bed before applying the next one). Rinse the beaker
with three 2 ml portions of 2M NH_4SCN solution and put the rinses on the
column as for the main solution.
<u>3-3</u>. Elute the lanthanons at a flow-rate of 0.7 - 1.2 ml/minute with 150
ml of 2M NH_4SCN/0.2% $NH_2OH \cdot HCl$. (Note 3-b)
<u>3-4</u>. Elute the Am with 70 ml 4M HCl at the same flow-rate. (Note 3-c).
<u>3-5</u>. Collect the Am fraction in a glass beaker containing 15 ml 16M HNO_3
which is continuously stirred (magnetic stirrer).
<u>3-6</u>. Evaporate this fraction until only H_2SO_4 remains and then heat strongly
until the H_2SO_4 is volatilised (Note 3-d).
<u>3-7</u>. Wash beaker well with 12 M hydrochloric acid and evaporate again to
dryness.
<u>3-8</u>. Repeat steps 3-1 to 3-5. (Note 3-e).

NOTES

<u>3-a</u>. Dimensions of column: 1 cm x 20 cm. Resin: AG 1 x 8, (BioRad Lab-
oratories, Richland, California). Mesh size 100-200. Conditioned by passing 200 ml 4M HCl, 100 ml 0.01M HCl, 50 ml 2M NH_4SCN at a flow-rate of <u>0.5 -
1.0 ml/minute</u>.
<u>3-b</u>. If convenient, this may be done overnight. The NH_2OH maintains any
remaining Ce in the +3 form. If any is present as +4, it contaminates the
Am fraction and is electrodeposited with Am.
<u>3-c</u>. Column regeneration: May be made by washing with 200 ml of water con-
taining 2 drops of 12 M HCl, followed by 50 ml of 2M NH_4SCN.
<u>3-d</u>. The H_2SO_4 is produced by the oxidation of thiocyanate by HNO_3/HCl.
This oxidation proceeds smoothly when the reaction is carried out during
elution of the Am fraction. If it is made <u>after</u> collection of the Am
fraction, it should be carried out with care, adding small portions of the
Am fraction to the HNO_3 as the reaction is very vigorous -- also there is a
time lapse in the start of the oxidation.

<u>3-e.</u> This second thiocyanate column is essential to remove traces of the
<u>Nd</u> carrier which are still present after only one thiocyanate column sepa-
ration and are sufficient to degrade the alpha-spectrum.

4. ELECTRODEPOSITION OF SAMPLE.

<u>4-1.</u> Remove the stirring bar and evaporate the solution until all HCl and
HNO_3 are driven off.
<u>4-2.</u> Destroy any ammonium salts that are present by washing the beaker with
<u>12M</u> HCl and 16M HNO_3 and evaporating until only H_2SO_4 is left refluxing in
the beaker. (Note 4-a).
<u>4-3.</u> After cooling, wash beaker and beaker cover with water and continue
evaporation cycle again until only about 0.1-0.2 ml H_2SO_4 remains.
<u>4-4.</u> Add exactly 5 ml water and use to wash beaker walls well.
<u>4-5.</u> Add a few drops of thymol blue and 0.6-0.8 ml of 14M NH4OH. Neutralize
with 18M H_2SO_4 and add one drop in excess of the neutral point. (Note 4-b).
<u>4-6.</u> Transfer the solution with washing (using 2 aliquots, each 2 ml, of
approximately 0.02M H_2SO_4) to a plating cell. (Note 4-c).
<u>4-7.</u> Add 1 drop of 2M chelating agent in ammonium form (either diethylene-
triaminepenta-acetic acid (DTPA), 1,2- diaminocyclohexane-NNN'N'-tetra-
acetic acid (CDTA) or EDTA in that order of preference), mix well and adjust
pH to pH 2-3 using 2M NH_4OH. The final volume should be 10 ± 1 ml and equal
in all samples plated in parallel.
<u>4-8.</u> Plate for 2 hours at 1 amp per cell. Finally add 1 ml 14M NH_4OH to
the cell, mix and switch off current. (Note 4-d).
<u>4-9.</u> Disassemble the cell, wash the plate with distilled water and acetone,
dry and store for counting.

NOTES

<u>4-a.</u> The amounts of ammonium salts left in the beaker depend on the rela-
tive amounts of SCN^-, HCl and HNO_3. Their destruction by evaporation with
HNO_3 or HCl or both is important. Indication of NH_4NO_3, is the production of
brown fumes of nitrogen oxides on addition of 12M HCl during evaporation.
If there is no indication of NH_4NO_3, any salts are likely to be NH_4Cl.
Evaporate with HNO_3 to destroy this salt. The final product is H_2SO_4 which
should reflux at high temperature with fumes of SO_3.
<u>4-b.</u> The amount of 14M NH4OH used is not critical as long as at least 0.6
ml is used. It should be the same in all samples being plated together so
as to equalize salt concentrations in the electroplating cells.
<u>4-c.</u> For details of plating cell, see (5). Currently a Hewlett Packard
DC Power Supply #6286A with variable output 0-20V, 0-10A is in use. It is
being used to plate 8 samples simultaneously in parallel. The 8 samples
should contain about the same salt concentration so that equal current
flows through each cell.
<u>4-d.</u> Shorter plating time may not give quantitative plating of Am. Longer
plating than 2 hours is not harmful. The solution must be neutralized before cutting off the current, to preclude re-solution of the sample.

RESULTS

In Tables 1, 2, and 3 are shown some data obtained using this procedure on biological and sediment samples. Also shown are the data for
$^{239,240}Pu$ determined on the same samples. Reproducibility appears quite
satisfactory from the data on samples analyzed in replicate. While the
ratios of ^{241}Am to $^{239,240}Pu$ appear not unreasonable, there are very few
other data with which they can be compared. The contrast between the
ratios found in Fucus (0.064) and in Sargassum (0.045), and those we have
seen (Tables 1 and 3) in sediments contaminated principally by fallout
(averaging only a little under 0.2) indicates to us that these seaweeds may
discriminate between these two radioelements. The IAEA Fucus sample was

TABLE I. REPRODUCIBILITY OF [241]Am AND [239,240]Pu ANALYSES OF SEDIMENTS

Sediment	Replicate	[241]Am d.p.m./Kg Dry Sediment		[239,240]Pu	
IAEA Intercomparison Sediment Sample #SD-B-1	1	470	± 20	1920	± 60
	2	470	± 20	1960	± 80
	3	510	± 20	1910	± 90
	4	530	± 20	2060	± 80
	5	570	± 20	1960	± 80
	6	520	± 20	2060	± 70
Buzzards Bay, Mass. June 1972, 0-1 cm.	1	29.6 ±	2.0	146	± 1
	2	30.5 ±	1.2	152	± 4
Lake Ontario August 1971	1	<2		12.8 ±	1.3
	2	<0.7		12.3 ±	1.1
	3	---		9.8 ±	1.0
	4	<2		11.2 ±	0.9

TABLE II. [241]Am AND [239,240]Pu ANALYSES OF SEAWEEDS

	Replicate	[241]Am d.p.m./1 g	[239,240]Pu (Dry Weight)
Fucus vesiculosus (IAEA Intercomparison Sample #AG-1-1)	1	12.0 ± 0.6	69.4 ± 4.8
	2	12.0 ± 0.6	67.1 ± 5.7
	3	10.5 ± 0.5	66.9 ± 3.4
	4	11.6 ± 0.6	56.5 ± 2.8
		d.p.m./Kg (Dry Weight)	
Fucus vesiculosus Plymouth, Mass., May 1972	1	0.5 ± 0.1	7.8 ± 1.2
Sargassum sp. ATLANTIS II, Cruise 20 Station 973, April 1966	1	10 ± 2	218 ± 4
	2	9.2 ± 0.9	212 ± 4

TABLE III. ^{241}Am/ ^{239,240}Pu RATIOS IN SEDIMENTS

Sediment	^{241}Am d.p.m./Kg	^{239,240}Pu Dry Sediment	^{241}Am/^{239,240}Pu
Buzzards Bay, Mass.			
May 1972			
0- 2 cm.	30.5 ± 1.2	152 ± 4	0.20 ± 0.01
2- 4 cm.	19.1 ± 1.0	90.7 ± 2.7	0.21 ± 0.01
4- 6 cm.	13.0 ± 0.8	66.8 ± 2.4	0.19 ± 0.01
6- 8 cm.	8.6 ± 0.6	43.1 ± 1.8	0.20 ± 0.02
8-10 cm.	8.1 ± 0.5	42.8 ± 1.8	0.19 ± 0.02
10-12 cm.	3.6 ± 0.4	25.4 ± 1.6	0.14 ± 0.02
12-14 cm.	2.0 ± 0.3	7.6 ± 0.9	0.26 ± 0.05
14-16 cm.	0.6 ± 0.1	3.7 ± 0.5	0.16 ± 0.03
Eel Pond, Woods Hole, Mass. **May 1972**	0.33±0.07	2.6 ± 0.2	0.13 ± 0.03
Lake Ontario, Ontario Basin **43°24'00"N, 79°00'36"W** **Depth 90 m.**			
0-2 cm.	74 ± 4	460 ±19	0.16 ± 0.01
2-3 cm.	31 ± 3	194 ±20	0.16 ± 0.02
3-4 cm.	16 ± 3	109 ± 8	0.15 ± 0.03
4-5 cm.	13 ± 1	40 ± 4	0.33 ± 0.04

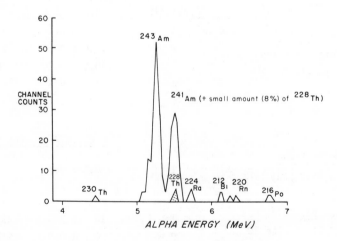

FIG. 1.   Alpha spectrum of Am fraction separated from 50 g sediment (Buzzards Bay, Mass., June 1972, 0-1 cm).

contaminated in a waste disposal situation; it is our expectation that the results of this intercomparison exercise are discussed elsewhere in this report.

In Figure 1 is shown the alpha spectrum obtained from one of the samples represented in Table 1. Despite rather low recovery (23% of the ^{243}Am tracer used to monitor yield), the modest counting period of 958 minutes was sufficient to give ample sensitivity for the ^{241}Am measurement on the 50 g sample used. The spectrum also illustrates the importance of thorium removal: The alpha particles from ^{228}Th are incompletely resolved from those of ^{241}Am, so that presence of a residue of ^{228}Th reduces the sensitivity of detection of ^{241}Am. As is illustrated here, when the amount of ^{228}Th is relatively small, its contribution to the ^{241}Am peak can be calculated in proportion to the well-resolved peak of the ^{228}Ra daughter. Such need for correction should, however, be minimized, since it always increases the uncertainty of the final value for ^{241}Am.

^{237}Neptunium analysis

In our separation scheme for Am and Pu analysis, Np is adsorbed with Pu from 8M HNO_3 on an anion resin. It is eluted with 4M HCl following Pu elution with 12M HCl/0.1 M NH_4I. We have been evaluating ^{239}Np derived from ^{243}Am decay as a Np tracer in the system.

The amount of ^{239}Np added with 1-2 d.p.m. ^{243}Am is inconveniently low for measurement using low background counters. However, 1000 d.p.m. ^{243}Am absorbed on a cation exchange resin contains enough ^{239}Np at equilibrium to provide sufficient activity for β or γ determination of 4-5 samples. Elution of ^{239}Np at weekly intervals (using 0.5 M HCl after periodic acid oxidation of Np) shows promise of providing a steady source of ^{239}Np tracer. The ^{239}Pu added to a sample by this procedure has been calculated to be about 0.0001 d.p.m. -- insignificant compared with Pu analytical sensitivity.

Our preliminary experiments with this system indicate that Np can be separated and electroplated with adequate tracer recovery. However, decontamination from U derived from environmental samples can prove troublesome. ^{237}Np alpha's (4.78 MeV, 75%; 4.65 MeV, 12%) are not resolvable from ^{234}U (4.77 MeV, 72%; 4.72 MeV, 28%).

ACKNOWLEDGEMENTS

Dr. C. Gatrousis, Lawrence Livermore Laboratory, California, kindly provided us with ^{243}Am tracer. Dr. V. T. Bowen and Dr. V. E. Noshkin started and encouraged the project. Our appreciation to the above is acknowledged with thanks.

The work was supported at Woods Hole Oceanographic Institution by the U. S. Atomic Energy Commission under contracts AT(11-1)-3563 and AT(11-1)-3568; R. Bojanowski worked at Woods Hole Oceanographic Institution on a UNESCO fellowship. This support is gratefully acknowledged.

This is Contribution No. 2970 from the Woods Hole Oceanographic Institution.

REFERENCES

[1] WONG, K. M., NOSHKIN, V. E., SURPRENANT, L., BOWEN, V. T., Plutonium 239 in some marine organisms and sediments, Health and Safety Laboratory Fallout Program Quarterly Summary Report HASL-227 (1970), I-25 to I-33.
[2] BOWEN, V. T., WONG, K. M., NOSHKIN, V. E., Plutonium-239 in and over the Atlantic Ocean, Journal of Marine Research 29 1 (1971) 1-10.

[3] WONG, K. M., BURKE, J. C., BOWEN, V. T., "Plutonium concentration in
    organisms of the Atlantic Ocean", Health Physics Aspects of Nuclear
    Facility Siting 2 (1971) 529-539.
[4] NOSHKIN, V. E., BOWEN, V. T., "Concentrations and distributions of
    long-lived fallout radionuclides in open-ocean sediments",
    Radioactive Contamination of the Marine Environment, IAEA
    (Vienna) (1973) 671-686.
[5] WONG, K. M., Radiochemical determination of plutonium in sea water,
    sediments, and marine organisms, Anal. Chim. Acta 56 (1971) 355-364.
[6] WONG, K. M., NOSHKIN, V. E., BOWEN, V. T., Radiochemical procedures now
    used at WHOI for the analysis of strontium, antimony, rare-earths,
    caesium and plutonium in sea-water samples, Reference Methods for
    Marine Radioactivity Studies, IAEA, Vienna (1970) 119-127.
[7] LIVINGSTON, H. D., MANN, D. R., BOWEN, V. T., "Double tracer studies
    to optimize conditions for the radiochemical separation of plutonium
    from large volume sea-water samples", Reference Methods for Marine
    Radioactivity Studies -- Determination of Transuranic Elements,
    Radioruthenium and other Radionuclides in Marine Environmental Samples,
    30 Oct. - 3 Nov. 1972, IAEA (Vienna), to be published.
[8] GIBSON, W. M., The radiochemistry of lead, National Academy of Sciences-
    National Research Council, Nuclear Science Series NAS-NS 3040 (1961)
    66-69.

ISOTOPIC ANALYSIS OF PLUTONIUM IN ENVIRONMENTAL SAMPLES BY ISOTOPE DILUTION MASS SPECTROMETRY*

R.J. DUPZYK, R.D. CARVER, I.A. DUPZYK
Lawrence Livermore Laboratory,
Livermore, Calif.,
United States of America

Short Communication

We have conducted a series of experiments using standardized [239]Pu ranging from 60 to 0.006 dpm or 300 to 0.003 pg. For samples as low as 0.006 dpm the accuracy of mass-spectrometric determination is better than ±2%, while at the 0.0006 dpm level we are able to measure [239]Pu within a factor of two of the correct value.

The method has been applied to isotopic analysis of plutonium that was isolated from soils, air filters, seawater, marine organisms and urine. We have found that the origin and size of original sample have little bearing on the accuracy of mass analyses. The most important factor is chemical purity.

With careful attention to the 'cleanliness' of chemical operations, it is possible to obtain reliable and accurate results for samples containing less than 0.1 picogram of total plutonium.

Samples were prepared as follows: Alpha-spectroscopy counting plates were stripped with 12N HNO_3. The samples were spiked with a known amount of isotopically pure [242]Pu mass tracer. Any extraneous inorganic ions were removed from the plutonium by passing the solution through a Dowex-1 ion exchange (nitrate form) column and the Pu was eluted with 8N HCl-0.05 N HI. After evaporating the solution to dryness several times with 10N HCl to remove any nitrate and iodine, the samples were loaded on canoe-shaped rhenium mass-spectrometer filaments. Mass analyses were made using a surface ionization source and a high sensitivity-low background tandem 60° sector magnetic mass spectrometer. Isotope ratios relative to the [242]Pu were measured. The [240]Pu/[239]Pu and [241]Pu/[239]Pu atom ratios, the atoms of [239]Pu, [240]Pu and [241]Pu, and the total α dpm were then calculated by using the atoms of [242]Pu spike previously added to the sample as a reference.

To date we have analysed approximately 40 environmental Pu samples. The activity varied from ~0.01 to ~50 dpm while the [240]Pu/[239]Pu ratio ranged 0.05 to 0.27/litre. These samples included plutonium isolated from soils throughout the USA, as well as from Atlantic Ocean water and sediments. Although some of the samples contained reactor-produced Pu (as evidenced by the low [240]Pu/[239]Pu ratio), many were considered to have been deposited from atmospheric fall-out. For these samples we measured [240]Pu/[239]Pu ratios in the range of 0.10 to 0.27.

---

* This work was performed under the auspices of the US Atomic Energy Commission

PROCEDURE FOR PLUTONIUM ANALYSIS OF LARGE (100 g) SOIL AND SEDIMENT SAMPLES

J.W.T. MEADOWS, J.S. SCHWEIGER, B. MENDOZA, R. STONE
Lawrence Livermore Laboratory,
Livermore, Calif.,
United States of America

Abstract

PROCEDURE FOR PLUTONIUM ANALYSIS OF LARGE (100 g) SOIL AND SEDIMENT SAMPLES.
A method for the complete dissolution of large soil or sediment samples is described. This method is in routine usage at Lawrence Livermore Laboratory for the analysis of fall-out levels of Pu in soils and sediments. Intercomparison with partial dissolution (leach) techniques shows the complete dissolution method to be superior for the determination of plutonium in a wide variety of environmental samples.

PROCEDURE

A 100 gram portion of the soil or sediment is placed in a large (1 litre volume) platinum crucible. An appropriate amount of plutonium (^{242}Pu) tracer solution is added followed by 200 ml of concentrated HNO_3. The mixture is stirred continuously while slowly adding a 250 ml portion of concentrated HF. With siliceous samples the initial reaction will be exothermic and the mixture will boil vigorously. Following this initial reaction the mixture is placed on a hot plate and boiled to about one-half of the initial volume. An additional 250 ml of HF are added with stirring and the solution is heated again to near dryness.

About 500 ml of concentrated $HClO_4$ are added and the sample is evaporated to dryness, followed by continued heating until the residue forms a hard crust and perchlorate fumes are no longer evolved.

The residue is dissolved by boiling with 800 ml of $6\underline{M}$ HNO_3 and a few ml of 30% H_2O_2.

Any undissolved residue is separated by centrifugation and retreated with HF, then $HClO_4$, and again baked to dryness. Most of this residue is dissolved by boiling with $6\underline{M}$ HNO_3 and treating with a few ml of 30% H_2O_2.

Any remaining residue is collected by centrifugation, transferred to a 50 ml iron crucible and dried. This residue is fused with a few grams of NaOH and cooled. The fused mass is dissolved by leaching the crucible with $6\underline{M}$ HNO_3. The undissolved residue after this treatment is quite small and is not treated further since no plutonium has been detected in material surviving the vigorous dissolution treatment.

The combined solution is evaporated to about 1000 ml volume and the acidity is adjusted to $7 - 8\underline{M}$ in HNO_3. A few grams of $NaNO_2$ are added to the hot solution to stabilize Pu in the IV valence state. The solution is cooled and passed through a column of Dowex 1×8 anion resin (6 in by 1 3/8 in dia., 50 - 100 mesh) previously equilibrated with $8\underline{M}$ HNO_3. The column is washed thoroughly with about 400 ml $8\underline{M}$ HNO_3, followed by about 400 ml $10\underline{M}$ HCl.

The Pu is eluted from the column with about 500 ml of $10\underline{M}$ HCl-$0.5\underline{M}$ HI mixture. The eluant is then evaporated to near dryness, a few ml of HNO_3 are added to destroy free I_2 and evaporation is continued. The sample is dissolved in dilute HNO_3, and after addition of lanthanium carrier, Pu is co-precipitated as the fluoride. The fluoride precipitate is dissolved in a mixture of saturated H_3BO_3 and HCl and treated with NH_4OH to precipitate the hydroxides. After centrifugation, the sample is dissolved in a few ml of concentrated HNl containing a few drops of concentrated HNO_3.

The Pu is again adsorbed on anion resin (chloride form) using a column of 8 cm by 6 mm diameter. The column is washed with about 15 ml of $10\underline{M}$ HCl and the Pu eluted with about 15 ml of $10\underline{M}$ HCl - $0.5\underline{M}$ HI. The eluant is then evaporated to near dryness, a few ml of HNO_3 are added to destroy free I_2 and evaporation is continued. A few ml of concentrated H_2SO_4 are added and the solution is evaporated to less than one ml volume. The pH is adjusted to about pH 4 - 5 with NH_4OH using methyl red as the indicator.

The solution is transferred to a plating cell and Pu is electrodeposited on a clean platinum counting disc using a current of 0.6 amp for about $1\frac{1}{2}$ to 2 hours. Immediately prior to shutting off current an excess of HN_4OH is added to avoid re-solution of the deposited Pu. The platinum disc is rinsed with water, methanol and dried. The plutonium concentration is then determined by alpha pulse-height analysis.

Data obtained by this technique were compared with results obtained by use of a leach technique.[1] For most of the samples that we have analysed with both techniques, the results using complete dissolution were 30 - 50% higher than those using the leach method. In all cases the leach residues were completely dissolved and analysed for plutonium to ensure that the discrepancies were not due to inhomogeneities in the sample. Although the leach method did compare favourably with complete dissolution for a number of samples, the variation in soil characteristics, the chemical form of plutonium, and other factors that determine the extent to which plutonium is bound to the matrix material do not allow us to predict with assurance the efficiency of leaching. Consequently, we have chosen to completely dissolve all soil and sediment samples for plutonium analysis.

---

[1] HARLEY, J.H., Ed., Manual of Standard Procedures, Health and Safety Laboratory, USAEC Rep. NYO-4700 (1970).

RADIOCHEMICAL DETERMINATION OF PLUTONIUM IN MARINE SAMPLES BY ION EXCHANGE AND SOLVENT EXTRACTION

A. AARKROG
Health Physics Department,
Danish Atomic Energy Commission,
Research Establishment Risø,
Roskilde, Denmark

Abstract

RADIOCHEMICAL DETERMINATION OF PLUTONIUM IN MARINE SAMPLES BY ION EXCHANGE AND SOLVENT EXTRACTION.

The solid samples are ashed at 600°C, and ^{236}Pu spike and Fe scavenger are added to 2 - 10 g of ash for one analysis. Potassium pyrosulphate is mixed with the ash in the ratio 3 : 1, and the mixture is heated to melting. By this treatment the oxidation step of Pu is adjusted to 4+ and all Pu brought into a soluble form. The Pu is precipitated as hydroxide along with the Fe scavenger, and the mixture of Fe and Pu is ion exchanged on an anion resin, which separates Fe from Pu. Finally the Pu solution is subjected to extraction with 1-phenyl-3 methyl-4-benzoyl-pyrazolone-5 in xylene. The organic phase, which contains the plutonium, is evaporated ona stainless steel planchet, ignited and counted on a silicon-surface barrier counter connected with a multichannel analyser.

INTRODUCTION

Plutonium metal oxidizes readily in air, particularly in the presence of moisture. The predominant oxide formed is PuO_2. After a non-critical destruction of a nuclear warhead the environmental contamination with plutonium will thus be in the form of plutonium oxide particles.

This was also the case in Thule, Greenland, where such a contamination occurred in 1968. The plutonium oxide was contained in a mixture of jet fuel, soot, silicone oil, and minute fragments of plastics and insulation materials from the aircraft. Plutonium oxide is greatly insoluble in water and dilutes acids, and we therefore found it justified to apply a potassium pyrosulphate fusion to make sure that all plutonium was brought into a soluble form.

The $K_2S_2O_7$ fusion adjusts the oxidation steps to tetravalent plutonium. However, the sulphate ions from the fusion cause trouble because tetravalent plutonium forms fairly stable sulphate complexes, e.g. $Pu(SO_4)_3^{2-}$. These complexes prevent an organic complex such as $Pu(TTA)_4$ from being formed. This meant that the conventional extraction method with thenoyltrifluoroacetone (TTA) was out of question.

THEORY AND EXPERIMENTS

During the work with extraction methods for radionuclides the Chemistry Department at Risø [1 - 4] has found that 1-phenyl-3 methyl-4 acyl-pyrazolones are able to extract tetravalent plutonium from strong nitric acid solutions.

91

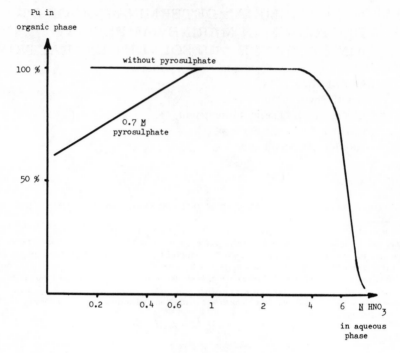

FIG.1.   Extraction of Pu with BP (1-phenyl-3 methyl-4-benzoyl-pyrazolone-5) for aqueous solution with
(0.7M) and without pyrosulphate.

1-phenyl-3 methyl-4 acyl-pyrazolones are β-diketones and may therefore be present in two tautomeric forms, the keto and the enol forms:

$$\text{enol form (a)} \rightleftharpoons \text{keto form} \rightleftharpoons \text{enol form (b)} \tag{1}$$

The enol form (HP) is a fairly strong acid and reacts with tetravalent
plutonium:

$$Pu^{4+} + 4\,HP \rightleftharpoons PuP_4 + 4\,H^+ \tag{2}$$

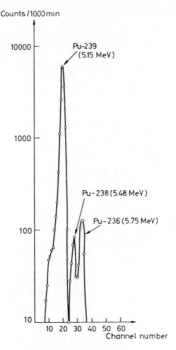

FIG.2. Alpha spectrum from a bivalve sample collected at Thule in August 1968.

PuP_4 is an organic complex and is easily extracted by organic media, e.g. xylene, which is used for the dissolving of the pyrazolone.

The sulphate ion has a depressive effect on the extraction because inorganic plutonium complexes are formed according to

$$Pu^{4+} + n\,SO_4^{2-} \rightleftarrows Pu(SO_4)n^{4-2n} \qquad (3)$$

these complexes are hydrophilic and thus not extractable into the organic phase. However, if the acid concentration is sufficiently high (1-6N HNO_3), the equilibrium

$$HSO_4^- \rightleftarrows SO_4^{2-} + H^+ \qquad (4)$$

is displaced to the left and the sulphate complexes in (3) are not formed. On the other hand, (2) shows that at increasing acidity this equilibrium is displaced to the left and the PuP_4 complex formation is obstructed.

Figure 1 illustrates the plutonium extraction with 1 phenyl-3 methyl-4-benzoyl-pyrazolone-5 (BP). The depressing effect of sulphate (from the $K_2S_2O_7$ fusion) is perceptible up to 1N HNO_3. From 4N HNO_3 the efficiency of the BP is rapidly decreasing according to (2).

The optimum region for plutonium extraction with BP is thus 1-4N HNO_3. In this interval neither Am^{3+} nor Th^{4+} or UO_2^{2+} were extracted by BP. For Am^{3+} we measured a decontamination factor of 10^5.

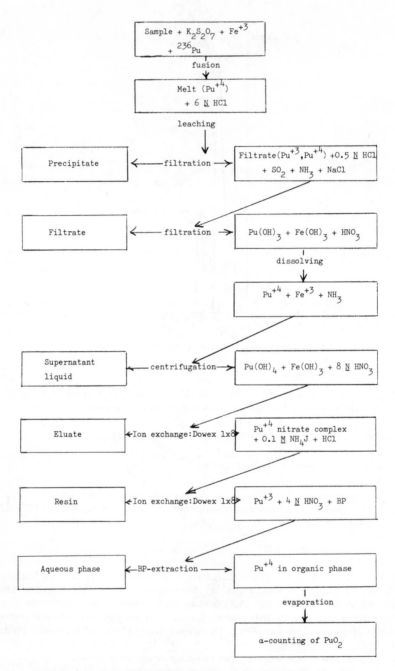

FIG.3.   Outline of the procedure for an ashed sample.

PRACTICAL APPLICATION

The pyrosulphate fusion followed by the BP-extraction method was applicable to snow and ice samples collected after the Thule accident. If, however, the samples contained iron, we ran into difficulties because BP forms an iron pyrazolone complex.

From the BP-extraction we got our counting sample directly from an evaporation of the xylene phase with the plutonium pyrazolone complex, and, as shown in Fig.2, this gave us reasonably thin samples for α-spectroscopy. Such samples were not obtainable if iron was present. We therefore found it necessary for most environmental samples to combine the extraction method with an anion exchange procedure, which we adapted from Bowen [5].

The principle in the anion exchange of plutonium is based on the fact that tetravalent plutonium easily forms negatively charged nitrate complexes, which disintegrate by reduction of the plutonium to the trivalent stage.

In the present procedure the anion exchange resin (Dowex 1×8, 50-100 mesh) is transformed to the nitrate form and the analysis in $8\underline{N}$ HNO_3 is transferred to the column. The plutonium complex $(Pu(NO_3)_6^{2-}$ is retained on the resin, while Fe^{3+} is eluated with portions of $8\underline{N}$ HNO_3. The column is percolated with hydrochloric acid, and the plutonium is eluted from the column with NH_4I in hydrochloric acid. The eluate is evaporated with HNO_3 and the solution is made $4\underline{N}$ with regard to HNO_3. This solution is extracted with BP, the aqueous phase is discarded, and the organic phase, which contains the plutonium, is evaporated on a steel planchet, ignited and counted on a silicon surface barrier α-counter connected to a multichannel analyser. Figure 3 shows a flow sheet of the total procedure [6] for solid samples. In the case of seawater the first step is a coprecipitation with ferrous hydroxide with conc. NH_4OH.

DISCUSSION

The procedure requires 2 to 3 days. It separates plutonium from other α-emitters (Am, Th, U, Po) and it normally gives thin samples, which permit a differentiation between ^{239}Pu, ^{238}Pu and ^{236}Pu. The procedure has been used for marine samples collected at Thule [7].

However, the yields have not been satisfactory. Sediments showed an average yield of $39\pm1\%$ (± 1 SE), seawater $44\pm 6\%$, and soft tissue of bilvalves $54\pm 3\%$. The sensitivity of the procedure, defined as three times the activity found in a blank, was better than 1 fCi/l seawater for a 50 litre sample and approximately 10 fCi/g ash for a 4 g ashed sample. While these sensitivities were sufficient for the Thule samples, they are not always satisfactory for fall-out background measurements.

On account of great inhomogeneities in the Thule samples, it was difficult to get a realistic estimate of the reproducibility of the procedure. The relative error of 21 sets of replicates of seawater, sediments and organisms varied from 4 to 100% with a median value of 26%. As expected, we found the greatest deviations in the samples collected at the bottom, as these samples were especially susceptible to contamination with 'hot' particles.

REFERENCES

[1] JENSEN, B.S., Acta Chem. Scand. $\underline{18}$ (1964) 739-49.
[2] JENSEN, B.S., Acta Chem. Scand. $\underline{13}$ (1959) 1347-57.
[3] JENSEN, B.S., Acta Chem. Scand. $\underline{13}$ (1959) 1668-70.
[4] JENSEN, B.S., Acta Chem. Scand. $\underline{13}$ (1959) 1890-96.
[5] BOWEN, V.T., personal communication, 1968.
[6] MARKUSSEN, E.K., Rep. Risø-M-1242 (1970).
[7] AARKROG, A., Health Phys. $\underline{20}$ (1971) 31-47.

DETERMINATION OF PLUTONIUM
IN THE MARINE ENVIRONMENT

K.C. PILLAI
Health Physics Division,
Bhabha Atomic Research Centre,
Trombay, Bombay
India

Abstract

DETERMINATION OF PLUTONIUM IN THE MARINE ENVIRONMENT.
Analytical methods used in the measurement of plutonium in environmental samples are discussed. Data is present from field studies carried out in the coastal region near the Trombay discharge area (Bombay, India) of strontium-90, ruthenium-106, caesium-137, cerium-144 and plutonium-239 on samples of seawater sediments and biological material. Dose rates to organisms for these radionuclides are calculated.

INTRODUCTION

Plutonium is the first artificially produced element that has found extensive use in nuclear energy programmes. It has been also used as a power source in spacecrafts, for cardiac pacemakers etc. In view of the establishment of many irradiated fuel reprocessing plants in different parts of the world and the subsequent radioactive waste releases from them and the increasing trend in use of plutonium the possibility of more plutonium entering the environment has increased.

It was reported earlier that plutonium exists in nature in very small quantities in uranium-rich minerals [1,2]. However, the element as a contaminant entered the environment as a result of atomic weapon tests and measurable quantities of plutonium have appeared in the global fall-out. Varying estimates of plutonium fall-out have been reported [3-7]. In one recent report by Harley [8] plutonium fall-out has been estimated as 300 kCi from the beginning up to the test moratorium, with a probable addition of about 5% from subsequent tests by China and France.

The major isotope of plutonium produced and utilized is Pu-239. The isotopes of plutonium detected in fall-out are plutonium-239, 240, 241, 242 and 238. As a result of a satellite failure in 1964 about 17 kCi of ^{238}Pu (1 kg) was released and subsequently deposited on earth's surface. Other cases of accidental releases of plutonium into the environment have occurred at Spain, Thule, Greenland, Rocky Flats etc., about which full information is not available.

PLUTONIUM MEASUREMENTS IN THE MARINE ENVIRONMENT

The oceans, by virtue of their larger surface area, receive a major share of fall-out plutonium. A fraction of the land run-off also reaches

the oceans. Many coastal installations that reprocess irradiated fuels
discharge their low-level radioactive liquid effluents into the sea. In spite
of the generally efficient recovery of plutonium in the reprocessing of
irradiated fuels, traces of plutonium can find their way into the environment through the waste releases from such plants.

Until 1964 there was no data available on the distribution of plutonium
in the marine environment. Pillai et al. [6] measured plutonium concentrations in some coastal and off-shore waters of the North Pacific and in
some coastal organisms. The seawaters showed plutonium levels of 10^{-4}
to 10^{-3} pCi/l and some of the organisms analysed showed high concentration
factors for plutonium. From these studies it was also possible to estimate
the ratio of $^{238}Pu/^{239}Pu$ in the marine environment prior to the deposition
of ^{238}Pu from satellite failure.

Further work [9] on the distribution of plutonium in two vertical
profiles (1000 m) from the Pacific Ocean during 1964 indicated that
plutonium has penetrated into the ocean depths. Subsequent studies on
plutonium content in some sediments from the Atlantic Ocean indicated
measurable quantities of plutonium in surface sediments as well.

Mikake et al. [10,11] reported that seawater samples from the Western
North Pacific during 1967 contained the same levels of plutonium as
obtained for coastal waters of the eastern North Pacific in 1964 (10^{-4} to
10^{-3} pCi/l). However, the $^{238}Pu/^{239}Pu$ ratio was reported to be higher. In
the summer of 1968 seawater samples collected from the Japan Sea showed
^{239}Pu concentrations of 10^{-4} - 10^{-3} pCi/l.

Bowen et al. [12] made measurements of plutonium concentrations in
surface and depth samples from the Atlantic Ocean during 1968-1969.
Surface samples showed concentration of 10^{-4} - 10^{-3} pCi/l of plutonium-239.
They obtained maximum values of ^{239}Pu at 500 m depth, which they atributed
to the biological sedimentation of plutonium. Wong et al. [13,14] obtained
high concentration of plutonium for organisms collected from Atlantic Ocean.
The results indicated high concentration of plutonium in some tissues —
liver, bone, gut — of each organism examined.

Aarkrog [15] and Hanson [16] investigated plutonium levels in the
Arctic environment subsequent to the release of plutonium (in 1968) on
the sea of Bylot Sound near Thule, Greenland.

The above studies indicated the levels of plutonium in the marine
environment. Further studies are required to fully understand the physico-
chemical, biological and geochemical behaviour of plutonium in the marine
environment. Most of the measurements of plutonium reported are for the
open ocean and information on the distribution of plutonium in the coastal
environment is not available.

METHODS FOR THE DETERMINATION OF PLUTONIUM IN MARINE
SAMPLES

The radiochemical separation of minute traces of plutonium from the
whole spectrum of elements in seawater is difficult. Since plutonium exists
in a multiplicity of oxidation states with different chemical properties,
these are made use of in isolation and radiochemical separation of this
element. Most of the procedures involve coprecipitation, complexation,
solvent extraction and ion exchange.

TABLE I. RECOVERY OF PLUTONIUM IN MARINE SAMPLES BY
VARIOUS PROCEDURES

Sample	Procedure	Sample size	No. of samples	Recovery (%)	Remarks
	Seawater				
1.	BiPO$_4$ precipitation, LaF$_3$ and anion exchange [a]	45 litres	10	31.6 - 76.3	>50% recovery for 7 samples
		4 litres	2	60.2 - 63.7	
2.	Ferric hydroxide precipitation, LaF$_3$ and anion exchange [a]	200 litres	2	26 - 30.6	
		2 litres	1	65.0	
3.	Ferric hydroxide precipitation and anion exchange	30 litres	3	27.6 - 56.5	
	Supernatant from above, again ferric hydroxide precipitation and anion exchange		1	1.85	
	Organisms				
1.	Dry ashing at 600°C, dissolution in nitric acid, BiPO$_4$ precipitation, LaF$_3$ and anion exchange		10	50.2 - 74.6	>60% recovery for 9 samples
2.	Initial treatment with nitric acid, ashing at 400-450°C, BiPO$_4$ and anion exchange		11	26.9 - 98.7	>55% recovery for 9 samples
	Sediments				
1.	Extraction with 8N HNO$_3$, BiPO$_4$ precipitation, LaF$_3$ and anion exchange	5-43 g	12	28.4 - 87.7	>50% for 8 samples

[a] Work carried out at Scripps Institution of Oceanography, University of California, San Diego, USA.

Some preliminary concentration method, usually coprecipitation, is
used for isolating plutonium from the bulk of the sample. Kooi and co-
workers [17,18] used BiPO$_4$ for carrying down plutonium from seawater
samples. Pillai et al. [6,19] adopted this method for separating plutonium
from large volumes of seawater (45 litres). The BiPO$_4$ precipitate was
dissolved in HCl, Pu carried on lanthanum fluoride and subsequently separated
using an anion exchange column. Reasonable recoveries and absolute
radiochemical purity were obtained for this procedure. Higher recoveries
were obtained for lower volumes of samples. This procedure had the
advantage that initial precipitation was carried out in the acid condition
and hence there was very little chance of plutonium losses by container
absorption and polymer formation.

The marine organisms were dried and dry ashed at 600°C. The residue was dissolved in nitric acid and the solution, after proper dilution, was treated as in the case of seawater. The same method was followed for marine sediments after repeated extraction with $8\underline{N}$ HNO_3. For samples of marine organisms and sediments containing large amounts of iron this procedure completely removes the iron, ensuring a clear thin electroplated source for alpha-spectrometric measurements.

Subsequent studies in our laboratory have shown that for marine organisms a preliminary treatment with concentrated nitric acid, evaporating off the acid and heating in a muffle furnace at 400-450°C, give easily soluble residues.

The recovery results obtained for various procedures carried out for seawater, marine organisms and sediments are given in Table I.

Ferric hydroxide was used for isolating Pu from large volumes of seawater (200 litres). The precipitate was dissolved in hydrochloric acid, Pu was carried on LaF_3 and separated by anion exchange. Low recoveries were obtained for large volumes of samples (see Table I).

Coprecipitation with ferric hydroxide was made by Wong et al. [20] for isolating Pu from seawater (55 litres), organisms and sediments. The ferric hydroxide precipitate is dissolved in $8\underline{N}$ HNO_3, passed through an anion column, followed by washing with $8\underline{N}$ $H\overline{N}O_3$ and conc. HCl, and plutonium eluted by $HCl-NH_4I$ solution. By using this method Bowen et al.[12] reported very low recoveries for seawater (1-57%).

The above method was further modified by Wong et al. [21]. Instead of carrying Pu on ferric hydroxide, iron was reduced by addition of sodium sulphite and plutonium (III) was carried on ferrous hydroxide. The precipitate was dissolved in $8\underline{N}$ HNO_3 and Pu separated on an anion exchange column. The ion exchange was repeated on an anion column in HCl medium to remove final traces of Th. Recoveries of $52 \pm 18\%$ were reported for samples (5-60 litres) of seawater. Higher recoveries were attributed to the absence of Pu(IV) polymer by reduction of Pu to Pu(III). The marine organisms after ashing (450°C) were dissolved in nitric acid and plutonium purification was carried out by ion exchange. The sediments after extraction with HNO_3-HCl were directly passed through anion column. The ion exchange was repeated using an anion column in HCl medium for removal of Th. For sediments and organisms $63 \pm 20\%$ recovery were reported.

Aarkrog used potassium bisulphate for solubilizing Pu in biological ash samples. Pu was carried on ferric hydroxide and was separated by anion exchange on a nitrate column. The plutonium solution was subjected to extraction with 1 phenyl-3 methyl-4 benzoyl-pyrazolone-5 in xylene. The organic phase was evaporated to dryness and counted.

Ion exchange in $8\underline{N}$ nitric acid medium usually affects the column by gas formation, especially when the resin size used is smaller.

Other methods using 8-acetoxy quinoline [22], potassium rhodizonate [23], barium sulphate [24], cupferride [25] etc. are reported for plutonium estimation but are not suitable for large volumes of environmental samples.

Unlike other radionuclides which are commonly found in marine samples, plutonium estimations cannot be based on any chemical yield obtained in the procedure followed. None of the procedures mentioned above give consistent recoveries for plutonium. Therefore it is now an accepted practice to use an internal tracer for correcting the recovery obtained. ^{236}Pu is the tracer commonly employed. Alpha spectrometry is used for the final assessment of plutonium activity.

TABLE II. INTERACTION OF LOW-LEVEL EFFLUENT WITH SEAWATER

Effluent No.	Volume of effluent (ml)	Volume of seawater	Period of contact	Plutonium in solution total (dpm)	(dpm/l)
I	1	200	15 days	7.0	35.0
	1	200	15 days	8.6	43.0
	1	200	15 months	8.17	40.9
II	1	1000	7 days	10.0	10.0
	10	100	3 days	1.8	18.0
	5	100	1 day	1.8	18.0

TABLE III. DISTRIBUTION OF RADIONUCLIDES IN SEAWATER AND SUSPENDED SILT FROM AREAS NEAR TO RADIOACTIVE EFFLUENTS DISCHARGE LOCATION IN THE BOMBAY HARBOUR BAY

	^{90}Sr	^{106}Ru	^{137}Cs	^{144}Ce	^{239}Pu
Seawater (pCi/l)	25.35	1.44	21.61	0.62	0.0042
Silt (pCi/g dry)	0.42	8.61	23.97	68.8	0.553
K_d	1.66×10^1	6.98×10^3	1.57×10^3	1.1×10^5	1.3×10^5
Seawater (pCi/l)	153.6	53.08	121	ND	0.0195
Silt (pCi/g dry)	-	462.9	127.2	1855	0.9412
K_d	-	8.72×10^3	1.05×10^3	-	0.48×10^5
Seawater (pCi/l)	3.9	ND	6.55	ND	0.012

PLUTONIUM IN COASTAL WATERS

The nature of plutonium in the marine environment depends upon the physico-chemical nature of the introduced plutonium — through fall-out, waste releases, run-off from land and other sources, and the subsequent interactions that take place in the seawater medium. Coastal waters assume significance because of the large inventory of wastes remaining in such areas and the human utilization of coastal waters.

It is reported [26] that 65-70 μg of plutonium metal can be dissolved in a litre of seawater. However, preliminary studies conducted by inter-acting low-level effluents from fuel-reprocessing plants with seawater (filtered through 0.22 μm millipore filter paper) showed insignificantly small amounts of Pu remaining in solution (Table II).

It is quite likely that such low levels of dissolved Pu in seawater are a result of the removal of plutonium along with other fission products (especially Ce). An attempt was made to study the ion-exchange nature of the dissolved plutonium in seawater but consistent results could not be obtained. Both the cation and anion columns retain Pu, and the indications are that the anion column retains more plutonium from the seawater.

TABLE IV. DISTRIBUTION OF RADIONUCLIDES IN BOTTOM
SEDIMENTS: DISCHARGE LOCATION

Location	^{106}Ru	^{137}Cs	^{144}Ce	^{239}Pu
		(pCi/g dry)		
1	116.8	129.3	278.6	0.86
2	59.1	89.3	221.9	0.39
3	177.5	296.5	538.8	0.73
4	59.1	116.1	255.4	0.54
5	-	1551	313.1	0.56
6	3.7	135.7	-	0.42
7	-	125.4	97.5	0.27
8	-	189.3	391.3	0.49
9	37.6	105.2	193.5	0.78
10	62.4	149	339.5	0.60
11	89.0	129.3	234.6	0.47

FIELD STUDIES

Samples of seawater collected from the Bombay Harbour bay from
locations near the discharge area in Trombay were analysed for the
predominant radionuclides. After initial settling the suspended silt were
separated from seawater by filtration through Whatman No.42 filter paper.
Both the fractions were analysed separately. Table III gives the data
obtained. Very high K_d factors are obtained for Pu in silt. As expected,
only ^{144}Ce had comparable K_d values. Consequently only very low levels
of plutonium are found in seawater. Even near the discharge locations
the plutonium concentrations are only 10-100 times the fall-out levels
observed in other areas. These concentrations are likely to be further
depleted with time by contact with fresh silt.

Periodic examination of bottom sediment in the bay has revealed that
most of the plutonium gets accumulated in the discharge locations and its
close neighbourhood. Table IV gives the concentration of radionuclides
in bottom sediments collected from the discharge location. Table V gives
the distribution of radionuclides in sediments collected along the shore.
Because of the heavy silt load in the bay waters and its high K_d factors
for plutonium, the silt scavenges the plutonium, which thus gets localized
near the discharge area.

Americium could not be estimated in sediments since the radiochemically separated fraction did not give statistically significant count
rates. The americium activity is expected to be low since mostly low
burn-up fuels are processed in the reprocessing plant.

Several species of organisms obtained in the bay were carefully analysed
for their plutonium content. The concentrations of plutonium and other
radionuclides for different parts of the organisms are given in Table VI.

TABLE V. DISTRIBUTION OF RADIONUCLIDES IN SHORE SEDIMENTS

Date	Location	^{90}Sr	^{106}Ru	^{137}Cs (pCi/g dry)	^{144}Ce	^{239}Pu
70-01-03	TNJ	3.3	-	87.35	57.63	0.37
72-08-31	TNJ	-	35.6	52.25	50.71	0.694
71-03-09	TNJ	-	116	349.2	302.1	1.61
69-12-10	Cirus	-	3.0	1337	189.7	0.082
70-03-06	Cirus	0.8	3.8	658	-	-
69-12-10	TCB (Trombay end)	-	-	9.1	3.0	0.009
71-05-07	TCB (Trombay end)	-	60.3	107.6	163.2	0.72
69-12-10	Mahul	-	-	17.3	4.5	0.027

TABLE VI. ACCUMULATION OF RADIONUCLIDES IN COASTAL ORGANISMS FROM AREAS CLOSE TO DISCHARGE LOCATION

Organism	Parts analysed	^{106}Ru	^{137}Cs (pCi/g wet)	^{144}Ce	^{239}Pu	Remarks
Crab Seylla serrata	Flesh	1.44	9.2	1.23	ND	Sample size small
	Shell	ND	25.68	ND	ND	Sample size small
Seylla serrata	Flesh	5.35	3.7	4.54	0.0016	
	Bone	-	15.52	-	0.0025	
Seylla serrata	Flesh	24.7	15.5	7.6	-	
	Shell + bone	85.3	54.7	ND	0.0033	
Prawns			8.18		ND	Sample size small
Oysters	Flesh	13.68	8.11	6.9	0.0139	Contamination from sediment suspected
	Shell	25.3	ND	ND	0.0017	
Mackerel	Flesh	ND	0.023	ND	0.0005	
	Bone	-	-	-	0.0028	
Arius sp.	Flesh	ND	29.6	ND	0.0004	
	Bone	ND	62.6	ND	0.0060	
	Gut	230.5	76.2	33.7	0.0023	
Arius sp.	Bone	ND	40.1	ND	0.024	

These organisms were obtained from areas close to the discharge locations, especially the benthic organisms. No attempt was made to calculate the concentration factors for plutonium in organisms. But the levels of accumulation of Pu in these organisms near the discharge locations are about 10-100 times higher than the levels observed in organisms from other areas. The accumulations of ^{106}Ru, ^{144}Ce and ^{137}Cs are very high in the benthic organisms, but the levels of plutonium are relatively low. A comparison of the data on sediments and organisms shows that sediments play a major role in the removal of plutonium from seawaters. The biological uptake and transport may be insignificant in coastal areas.

DOSE TO THE ORGANISMS

The dose received by organisms from accumulation of ^{239}Pu, ^{137}Cs, ^{144}Ce, ^{106}Ru, ^{40}K etc. were computed. The dose received from 1 pCi of the nuclide uniformly distributed in the organism (for a typical case) calculates to (in mrad/a) 97 for ^{239}Pu, 20.25 for ^{144}Ce, 4.43 for ^{137}Cs, 20.54 for ^{106}Ru and 9.21 for ^{40}K.

The dose to the benthic organisms near the discharge location is mostly contributed by ^{144}Ce and ^{106}Ru. The dose from Pu is negligible in these cases. However, the dose from fall-out plutonium might be significant compared with the dose from fall-out ^{137}Cs, ^{90}Sr etc. for organisms in areas away from discharge locations.

ACKNOWLEDGEMENTS

The author acknowledges the valuable guidance given by Dr. A.K. Ganguly, Director, Chemical Group, BARC. Dr. H.L. Volchok, USAEC, Health and Safety Laboratory, New York, supplied ^{236}Pu standards and the author is grateful to him for the same. The assistance received from Shri A. Kuttappan in collection and processing of samples and Shri R.V. Pawaskar in alpha-spectrometric work is gratefully acknowledged.

REFERENCES

[1] GARNER, C.S., BONNER, N.A., SEABORG, G.T., Search for elements 94 and 93 in Nature. Presence of 94^{239} in Carnotite, J. Am. Chem. Soc. 70 (1948) 3453.

[2] CHERDYNTSEV. V.V., et al., ^{239}Pu in Nature, AEC. Tr-6968.

[3] STEWART, N.C., CROOKS, R.N., RISCHER, E.M.R., The Radiological Dose to Persons in the U.K. due to Debris from Nuclear Test Explosions prior to January 1956, Rep. AERE/HP/R-2017 (1956).

[4] LANGHAM, W., ANDERSON, E.C., Entry of Radioactive Fall-out into the Biosphere and Man, Rep. HASL-42 (1958) 282.

[5] OLAFSON, J.H., LARSON, K.H., "Plutonium, its biology and environmental persistence", Radioecology (SCHULTZ, V., KLEMENT, A.W., Jr., Eds), Reinhold, New York (1963) 633.

[6] PILLAI, K.C., SMITH, R.C., FOLSOM, T.R., Plutonium in the marine environment, Nature (London) 203 4945 (1964) 568.

[7] MIYAKE, Y., KATSURAGI, Y., SUGIMURA, Y., Depositions of plutonium in Tokyo through the end of 1968, Pap. Meteorol. Geophys. 19 2 (1968) 267.

[8] HARLEY, J.H., "Worldwide plutonium fallout from weapon tests", Environmental Plutonium Symp., LASL, Aug. 1971.

[9] PILLAI, K.C., SMITH, R.C., FOLSOM, T.R., Unpublished work, 1964.

[10] MIYAKE, Y., SUGIMURA, Y., Plutonium content in the Western North Pacific waters, Pap. Meteorol. Geophys. 19 3 (1968) 481.

[11] MIYAKE, Y., KATSURAGI, Y., SUGIMURA, Y., A study on plutonium fallout, J. Geophys. Res. 75 12 (1970) 2329.

[12] BOWEN, V.T., WONG, K.M., NOSHKIN, V.E., Plutonium-239 in and over the Atlantic ocean, J. Mar. Res. (1971).

[13] WONG, K.M., NOSHKIN, V.E., SUPRENANT, Lolita, BOWEN, V.T., "Plutonium-239 in some marine organisms and sediments", Health and Safety Laboratory, Fallout Studies Quarterly Summary Rep., HASL 227 (1970) 1-25.

[14] WONG, K.M., BURKE, J.C., BOWEN, V.T., Plutonium Concentration in Organisms of the Atlantic Ocean, Rep. NYO-2174-117 (1970).

[15] AARKROG, A., Radioecological investigation of plutonium in an Arctic marine environment, Health Phys. 20 (1971) 31.

[16] HANSON, W.C., Plutonium in lichen communities of the Thule, Greenland region during the summer of 1968, Health Phys. 22 (1972) 34.

[17] KOOI, J., HOLLSTEIN, U., An improved method for the determination of trace quantities of plutonium in aqueous media-1. Method and procedure, Health Phys. 8 (1962) 41.

[18] HOLLSTEIN, U., HOOGMA, A.H.M., KOOI, J., An improved method for the determination of plutonium in aqueous media-II. Interference of iron, calcium, uranium and chlorine; Application to sea water and urine, Health Phys. 8 (1962) 49.

[19] PILLAI, K.C., SMITH, R.C., FOLSOM, T.R., The separation and determination of plutonium in seawater and marine organisms, Environmental Marine Radioactivity Series, IMR-TR-922-66-B (1966).

[20] WONG, K.M., NOSHKIN, V.E., BOWEN, V.T., Reference Methods for Marine Radioactivity Studies, Technical Reports Series No. 118, IAEA, Vienna (1970) 119.

[21] WONG, K.M., Radiochemical determination of plutonium in sea water, sediments and marine organisms, Anal. Chim. Acta 56 (1971) 355.

[22] COLEMAN, G.H., The Radiochemistry of Plutonium, Rep. NAS-NS. 3058 (1965).

[23] SHIPMAN, W.H., WEISS, H.V., Rep. USNRDL-TR-451 (1960).

[24] SILL, C.W., WILLIAMS, R.L., Radiochemical determination of uranium and transuranium elements in process solutions and environmental samples, Anal. Chem. 41 (1969).

[25] BROOKS, R.O.R., Rep. AERE/AM-60.

[26] LINGREN, W.E., The electrochromatography of sea water containing dissolved plutonium, Rep. USNRDL-TRC-85 (1966).

RESULTS OF PLUTONIUM INTERCALIBRATION IN SEAWATER AND SEAWEED SAMPLES

R. FUKAI, C. N. MURRAY
International Laboratory of Marine Radioactivity,
Oceanographic Museum,
Principality of Monaco

Abstract

RESULTS OF PLUTONIUM INTERCALIBRATION IN SEAWATER AND SEAWEED SAMPLES.

The results of the intercalibration exercise for the measurement of plutonium-239 and 238 in two seawater samples SW-I-1 and SW-I-2 and a marine algae sample AG-I-1 are presented. Seventeen laboratories from 8 countries as well as the IAEA International Laboratory of Marine Radioactivity took part. A discussion of the results and methods used in the analysis is given. It is concluded that in spite of the complicated chemical procedures involved in plutonium analysis, the scatter of the reported results was much smaller than that for fission product radionuclides such as strontium-90, ruthenium-106, caesium-137 etc.

INTRODUCTION

During the IAEA Symposium on Rapid Methods for Measuring Radio-activity in the Environment held in July 1971 at Munich it became clear that controversial opinions exist among different laboratories concerning the suitable analytical procedures for transuranic elements, especially plutonium-238 and plutonium-239 [1]. Since these radionuclides with other transuranic isotopes might become important in the future from the stand-point of hazard assessment of the disposal of radioactive wastes from long-term operations of nuclear installations, and also since the analytical techniques for these elements constitute challenging problems in current radiochemical analysis, the intercalibration exercise specifically on analytical methods for transuranic elements was initiated by the Monaco Laboratory in 1971.

Laboratories experienced in the analysis of transuranic elements were invited to take part in this intercalibration exercise and two seawater samples SW-I-1 and SW-I-2 and one seaweed sample AG-I-1 were distributed for this purpose. These samples were known to be contaminated with the plutonium isotopes and other transuranic radionuclides under natural conditions in addition to fission product radionuclides at monitoring levels. Based on the results received by the Monaco Laboratory, a survey of the analyses is presented in this paper.

It should be emphasized that the execution of the intercalibration exercise was only possible with the voluntary contribution from the parti-cipating institutions of extra time and labour involved in the analytical work. The willing participation of many institutions in this exercise should be regarded as living proof or international collaboration within the scientific community.

COLLECTION AND PREPARATION OF SAMPLES

As has been discussed by Fukai et al. [2], considerable uncertainties in the chemical forms of radionuclides exist in environmental samples, so that the preparation of samples spiked with known amounts of radionuclides was purposely avoided. The field collection of the samples in late 1969 and early 1970 was carried out by the British Fisheries Radiobiological Laboratory, Lowestoft, with the financial support of the IAEA. For each collection 1000 litres of seawater were taken and stored in a plastic tank. The water was filtered through membrane filters of 0.45 μm pore size, homogenized thoroughly, acidified to pH 1.5 - 2 with hydrochloric acid, and then sub-sampled into 200 $\times$ 5 litre polyethylene containers. No special precautions were taken against contamination with trace metals.

The collection of the seaweed was carried out later, in early 1971. The samples were oven-dried at 100-105°C, pre-ground and pre-homogenized, and shipped to the Monaco Laboratory where final homogenizing and homogeneity tests were carried out.

HOMOGENEITY OF THE SAMPLES

As has been discussed in detail, the homogeneity of the seawater samples SW-I-1 and SW-I-2 with respect to various fission product radionuclides, excluding cerium-144, was estimated to be better than $\pm 3\%$ and $\pm 2\%$ respectively in terms of per-cent standard deviation of the content of the fission products concerned. The latter sample might show percentage standard deviation as high as ± 5 to 10% in rare cases. The results of the study on the stability of cerium-144 in a seawater medium (Preston and Dutton, unpublished data) showed that a considerable fraction of cerium-144 was deposited on the wall of polyethylene containers during a 6-month storage period, even when the pH value of the water samples was kept between 1.5 and 2. On the basis of these results, it can be considered that not only cerium-144 but also other lanthanide radionuclides are unstable in seawater medium, even if acidified, during prolonged storage. Although the atomic structure of actinides is similar to that of the lanthanides, the chemical properties of actinides are very much more variable, especially with respect to oxido-reduction states and complex formation. Nevertheless, there is a possibility that actinides, including plutonium isotopes, behave somewhat similarly to cerium-144 during storage. In fact, Laboratory Code No. 3 found that up to 20% of plutonium-238 and 239 in SW-I-2 were present on the wall of the container.

Homogeneity tests with respect to plutonium-239 carried out by Woods Hole Oceanographic Institution indicate, however, that the seawater samples used in this intercalibration exercise are reasonably homogeneous. The analytical results of plutonium-239 in samples SW-I-1 and SW-I-2 on a series of 4 and 6 separate bottles show percentage standard deviation for single determination of 11 and 6.8% respectively. Thus, considering the complexity of the procedures involved in plutonium-239 analysis, true homogeneity of these samples may be expected to be better than these values, possibly not significantly different from those estimated for the fission product radionuclides (excluding cerium-144).

TABLE I.   RESULTS OF ANALYSIS BY DIFFERENT INSTITUTIONS OF
SEAWATER SW-I-1 and SW-I-2

Code No.	SW-I-1		SW-I-2	
	^{238}Pu (10^{-2} pCi/kg)	^{239}Pu (10^{-2} pCi/kg)	^{238}Pu (10^{-2} pCi/kg)	^{239}Pu (10^{-2} pCi/kg)
1	---	5.4 ± 0.4	---	9 ± 3
2	0.75 ± 0.14	8.4 ± 0.4	2.2 ± 0.2	22.7 ± 1.1
3	0.7 ± 0.7[a]	8 ± 1[a]	4 ± 1[a]	28 ± 2[a]
4	1.1	10.0	2.2	16.9
5	2 ± 2	10 ± 2[a]	4 ± 2	20 ± 2[b]
6	0.7 ± 0.1	8.5 ± 0.4	2.6 ± 0.2	22.2 ± 0.6
7	0.89 ± 0.43[b]	9.2 ± 0.8[b]	3.27 ± 1.05	27.5 ± 2.5[b]
8	1.05 ± 0.58	6.90 ± 1.32	4.8 ± 1.4	23.9 ± 3.8
9	12 ± 10	77 ± 3	5.1 ± 0.1	34 ± 0.3
10	---	11.55 ± 1.59	---	25.38 ± 2.42
11	---	8 ± 4	---	13 ± 2
12	0.6 ± 0.1	10 ± 3[b]	3 ± 1	20 ± 2[b]

Values underlined are average values of 2 or more results.

[a] Errors represent 2 σ values.

[b] Values represent ^{239}Pu + ^{240}Pu.

The homogeneity of the seaweed sample AG-I-1 tested by gamma spectrometry was proved to be better than ±2%, in terms of per-cent standard
deviation of the content of fission product radionuclides, for a sample size
of 5 g.   Although Wong et al. [3] reported that plutonium-239 is concentrated
in the outer layers of giant brown algae Pelagophycus porra and that the
outside to inside ratio of plutonium-239 content reaches as high as 200, this
source of inhomogeneity should have been eliminated during the homogenizing
process.   Radionuclides such as zirconium-niobium-95 and cerium-144,
which are known to be concentrated in the surface layer of seaweeds, do not
show larger percentage standard deviation in comparison with potassium-40
or caesium-137, which have been shown to distribute themselves homogeneously in seaweeds.   Woods Hole Oceanographic Institution reported that
the percentage standard deviation for single determination of plutonium-239
for 6 different aliquots of 10 g dried seaweed was ±7.1%.   Thus, as in the
case of the seawater samples, true homogeneity may not be significantly
different from that estimated by gamma spectrometry.

PARTICIPATION

Sixteen institutions from 8 Member States of the IAEA reported their
results of analysis of plutonium.   A list of the institutions that participated
in this intercalibration exercise is given in Annex I.   The results presented
by these institutions are given under code numbers of the institutions.

TABLE II.   RESULTS OF ANALYSIS BY DIFFERENT INSTITUTIONS
ON SEAWEED AG-I-1

Code No.	^{238}Pu (pCi/g)	^{239}Pu (pCi/g)
1	2.7 ± 0.2	19.7 ± 0.5
2	3.9 ± 0.1	28.2 ± 0.3
3	4.1 ± 0.2[a]	27 ± 2[a]
4	4.09 ± 0.6	23.0 ± 2.5
5	3.9 ± 0.1	28.6 ± 0.3[b]
6	4.8 ± 0.2	30.0 ± 0.6
7	3.84 ± 0.07	26.7 ± 0.4[b]
8	4.17 ± 0.46	30.04 ± 3.00
9	3.7 ± 0.02	25.1 ± 0.03
10	6.97 ± 0.985	27.87 ± 0.27
11	3.1 ± 0.5	20.8 ± 3
12	---	28 ± 3
13	3.6 ± 0.4	24.3 ± 2.4
14	3.3 ± 0.1	22.3 ± 0.3
15	3.9 ± 0.3	27.3 ± 1.3
16	---	26.1 ± 2.2[a]
17	3.0 ± 0.6	24 ± 1[b]

Values underlined are average values of 2 or more results.
[a] Errors represent 2 σ values.
[b] Values represent ^{239}Pu + ^{240}Pu.

RESULTS REPORTED FOR PLUTONIUM-238 AND 239

The results of plutonium-238 and 239 measurements for seawater
samples SW-I-1 and SW-I-2 and seaweed AG-I-1 are given in Tables I and II
respectively. The values in these tables are presented as closely as
possible in original form reported by the participants; 1 July was taken as
reference date, and the units of the contents were normalized to pCi/kg and
pCi/g dried matter for seawater and seaweed respectively, and an average
value was presented when two or more values had been reported for one
sample. These averaged values are underlined in the tables. The estimated
errors given were also taken from or based on original reports. In cases
where 2 or 3 results were reported for one sample the errors for averaged
values were calculated from individual errors, while the standard deviations
of averaged values were calculated from individual results when four or
more results are available for one sample. The majority of these estimated
errors represent statistical error in alpha spectrometry.

RESULTS REPORTED FOR OTHER TRANSURANIC ELEMENTS

The following results were received on transuranic elements other than plutonium. Neptunium-237 was reported by a laboratory (Code No.6) for the seaweed sample as 0.013 ±0.003 pCi/g dried matter.

Americium-241 in the seaweed sample was reported by two laboratories (Code Nos 2 and 6) who used chemical separation techniques, their results being in each case 5.2 ±0.2 pCi/g dried matter. As can be seen, the values show amazing agreement. A further laboratory measured americium-241 directly using a Ge(Li) detector with a 3820-minute counting period. This result was 4.4 ±0.1 pCi/g and is very close to the alpha measurements of the other analyses.

One laboratory (Code No.2) measured two curium isotopes, curium-242 and 244, in the seaweed sample and gave values of 0.03 ±0.1 pCi/g dried matter for each isotope. This same laboratory also measured americium-241 and detected curium-242 and 244 in seawater SW-I-1. The result for the seawater for americium-241 was 0.13 ±0.03 pCi/kg. The ratio of americium-241 to plutonium-239 in seawater SW-I-2 is 0.57, while it is 0.20 in seaweed AG-I-1 − a not unreasonable agreement.

ANALYTICAL METHODS USED

Various analytical methods used by different laboratories are given in Annex II. Since small details of the analytical procedures might affect the overall performance of the methods, an attempt was made to describe the sequence of the procedures without omission of key procedures into a condensed form. To supplement the description in the table the literature for analytical method quoted by the participants is given in Annex III.

SURVEY OF THE RESULTS

A summary of the reported results is given in Table III. In this table maximum and minimum reported values are given with the ratios of maximum value to minimum value. These values do not include the values obtained for SW-I-1 by laboratory Code No.9, whose sample was thought to be contaminated on analysis. To obtain statistically significant average values of reported results a standard test of significance of these results was carried out by applying Chauvenet's criterion. On the basis of these tests it was found that only a few results lay outside the significant ranges. Specific information on each sample together with general remarks are given below. Although an attempt was made to correlate the quality of the values obtained with the method used, the information available is not sufficient to draw definite conclusions.

SW-I-1

Nine results were reported for plutonium-238, while twelve results were reported for plutonium-239. As mentioned above, the results from Laboratory No.9 for both plutonium-238 and 239 were higher than the other values reported by a factor of about 10. These values were thus excluded

TABLE III. STATISTICAL ANALYSIS OF RESULTS FOR SW-I-1, SW-I-2 AND AG-I-1

Sample	SW-I-1		SW-I-2		AG-I-1	
	^{238}Pu	^{239}Pu	^{238}Pu	^{239}Pu	^{238}Pu	^{239}Pu
No. of results considered	8	11	9	12	15	17
Maximum value (pCi/kg)	2×10^{-2}	11.55×10^{-2}	5.1×10^{-2}	34×10^{-2}	6.97×10^{3}	30.04×10^{3}
Minimum value (pCi/kg)	0.6×10^{-2}	5.4×10^{-2}	2.2×10^{-2}	9×10^{-2}	2.7×10^{3}	19.7×10^{3}
Max.-to-min. ratio	3.3	2.1	2.3	3.8	2.6	1.5
No. of results excluded[a]	1	0	0	0	3	3
Average (pCi/kg)	0.83×10^{-2}	8.7×10^{-2}	3.5×10^{-2}	22×10^{-2}	3.8×10^{3}	27.0×10^{3}
σ[b] (pCi/kg)	$\pm 0.07 \times 10^{-2}$	$\pm 0.5 \times 10^{-2}$	$\pm 0.4 \times 10^{-2}$	$\pm 2 \times 10^{-2}$	$\pm 0.1 \times 10^{3}$	$\pm 0.5 \times 10^{3}$
(%)	8.0	5.8	11	9.1	2.6	1.8

[a] Results excluded on the basis of Chauvenet's criterion.
[b] σ = standard deviation of the average.

TABLE IV.   RECENT DATA ON SW-I-2 FROM WOODS HOLE
OCEANOGRAPHIC INSTITUTION

Activity (pCi/kg × 10² at reference date 1 Jan. 1971)					
^{238}Pu	^{239,240}Pu	^{241}Pu	^{241}Am[a]	^{244}Cm	^{242}Cm[b]
2.9 ± 0.4	21.0 ± 1.4	--	9.1 ± 0.5	0.18 ± 0.05	< 6
3.2 ± 0.5	24.0 ± 1.8	--	10.2 ± 0.5	0.14 ± 0.05	< 9
3.5 ± 0.3	22.5 ± 0.9	721 ± 54	--	--	-

[a] These values have been corrected to 1 Jan. 1971 using the ^{241}Pu number. The measured values
(at 1 Nov. 1974) were quite a bit higher (13.1 and 14.1) from ingrowth of ^{241}Am from ^{241}Pu. A ^{241}Pu
half-life of 14.9 years seems to be the most recent and best value for this calculation.
[b] The ^{242}Cm numbers were calculated by correction of the Nov. 1974 data (< 0.014 and < 0.023 pCi/kg) to
1 January 1971 using a ^{242}Cm half-life of 163 days.

from the statistical considerations. Other results lie within the range of
0.6 - 2 × 10^{-2} pCi/kg for plutonium-238 and 5.4 - 11.6 × 10^{-2} pCi/kg for
plutonium-239 with the average value of 0.84 ± 0.07 × 10^{-2} pCi/kg and
8.7 ± 0.5 × 10^{-2} pCi/kg respectively. Although the per-cent standard
deviation of the average value for plutonium-238 is ± 8%, the estimated
errors for individual measurements presented in Table I show that many of
the measurements were made very close to the limit of determination.
Nevertheless, considering the difficulties involved in low-level radioactivity
measurements, the factor of 3 difference between maximum and minimum
values can be regarded as a reasonable agreement. In the case of
plutonium-239, whose level was one order of magnitude higher than that of
plutonium-238, the majority of the individual errors estimated were between
± 10% and ± 20%, and the percentage standard deviation for the average was
approximately 6%. The maximum-to-minimum ratio of 2.1 indicates better
agreement of the reported values than in the case of plutonium-238.

SW-I-2

In SW-I-2 the range and average value are respectively
2.2 - 5.1 × 10^{-2} pCi/kg and 3.5 ± 0.4 × 10^{-2} pCi/kg for plutonium-238, and
9 - 34 × 10^{-2} pCi/kg and 22 ± 2 × 10^{-2} pCi/kg for plutonium-239. It may be
significant that the maximum-to-minimum ratio of plutonium-239 for
SW-I-1 is much smaller than that for SW-I-2. Thus with the elevated level
of plutonium-239 the reported results seem to show more scatter. This
may be because many of the participating laboratories are engaged in
determinations of plutonium-239 in seawater at much lower levels, and the
working procedures established in the laboratories may not have always been
suitable for higher level samples.
New data for ^{238}Pu, ^{239,240}Pu, ^{241}Pu, ^{241}Am, ^{242}Cm, and ^{244}Cm in
SW-I-2 recently obtained by the Woods Hole Oceanographic Institution are
given in Table IV.

AG-I-1

Fifteen laboratories reported results for plutonium-238, while 17 reported results for plutonium-239. The range of reported values for plutonium-238 is 2.7 - 7 pCi/g with the average value of 3.8 ± 0.1 pCi/g, and that of plutonium-239 is 19.7 - 30.1 pCi/g with the average value of 27.0 ± 0.5 pCi/g. The maximum-to-minimum ratios for plutonium-238 and 239 are 2.7 and 1.5, respectively. Because of the relatively high levels of both plutonium-238 and 239 present the errors estimated for individual reported values rarely exceed $\pm 10\%$. Nevertheless, the scatter of the values for plutonium-238 in terms of maximum-to-minimum ratio is found to be similar to that of the seawater samples.

General Remarks

(1) In spite of the complicated chemical procedures involved in plutonium analysis, the scatter of the reported results was much smaller than that for fission product radionuclides such as strontium-90, ruthenium-106, caesium-137, etc. This may be because the laboratories that are capable of plutonium analysis maintain better house-keeping in radiochemical work and calibration of counting equipment.

(2) Many reported values showed that the concept of the 'significant figures' of analytical values was not taken into account in relation to estimated errors. While the estimated errors represent only counting statistics in many cases, the procedures of error estimations are uncertain in some cases. A guideline procedure for overall error estimation may be useful.

(3) Although a variety of analytical procedures were used by different laboratories, the principle of the analysis remains more or less similar in most of the methods used. With the exception of one method, all the procedures reported by laboratories for plutonium analysis used co-precipitation of iron hydroxide followed by separation of plutonium using anion exchange columns; after elution plutonium is electroplated usually onto stainless steel discs and alpha spectrometry is performed to determine plutonium-236 added as yield monitor as well as plutonium-238 and 239 present in the samples.

(4) The seaweed sample was found to contain more plutonium activity than was initially expected. Because of this increased activity there was a possibility that contamination of the surface barrier detectors could occur, as could cross-contamination of the other samples. A further possible difficulty was that the chemical yield measurements using plutonium-236 were open to greater error due to the grossly different activity levels of plutonium-236 added and plutonium-239 present in the samples.

(5) Certain calibrated plutonium-236 sources distributed in the recent past may be in error due to contamination with uranium-232 and thorium-228, which interfere with the plutonium-238 alpha spectrum. Great care must be exercised in the choice of calibrated standards so that such interferences are minimal.

(6) Some laboratories reported that difficulties occurred due to detectors being contaminated with americium-241 used by manufacturers for initial calibration. The possibility was raised of obtaining uncalibrated detectors from the manufacturers to reduce the risk of this type of contamination. It has been suggested (Livingston, private communication), how-

ever, that the contamination can be considered to result from recoiling uranium-232 arising from the use of too large an amount of plutonium-236 as yield monitor. The uranium-232 peak interferes with that of plutonium-238 in the alpha energy range around ~ 5.5 MeV.

(7) On the basis of the results obtained it seems logical to repeat this type of intercalibration exercise in the future with emphasis on the determination of americium-241 and other transuranic radionuclides.

REFERENCES

[1]   INTERNATIONAL ATOMIC ENERGY AGENCY, Rapid Methods for Measuring Radioactivity in the Environment (Proc. Symp. Neuherberg, 1971), IAEA, Vienna (1971) 942.

[2]   FUKAI, R., BALLESTRA, S., MURRAY, C.N., "Intercalibration of methods for measuring fission products in seawater samples", Radioactive Contamination of the Marine Environment (Proc. Symp. Seattle, 1972), IAEA, Vienna (1973) 3.

[3]   WONG, K.M., HODGE, V.F., FOLSOM, T.R., Plutonium and polonium inside giant brown algae, Nature (London) 237 (1972) 460-62.

ANNEX I

LIST OF INSTITUTIONS PARTICIPATING
IN THE ENVIRONMENTAL PLUTONIUM ANALYSIS

Country or Organization	Institution	Investigator
Australia	Australian Atomic Energy Commission, Research Establishment, Sutherland, N.S.W.	W.W. Flynn
Denmark	Danish Atomic Energy Commission, Research Establishment Risø, Health Physics Department, Roskilde	A. Aarkrog
India	Bhabha Atomic Research Centre, Health Physics Division, Trombay. Bombay 85	K.C. Pillai
Italy	Euratom-CCR, Ispra (Varese), Protection Division, Site Survey and Meteorology Section	P. Gaglione
Japan	Kanazawa University, Faculty of Science, Department of Chemistry, Kanazawa	M. Sakanoue
	Power Reactor and Nuclear Fuel Development Corporation, Tokai Branch, Analytical Chemistry Section, Tokai-Mura, Ibaraki-ken	T. Nishiya
Poland	Polish Academy of Sciences, Geophysics Department, Marine Station, Sopot	R. Bojonowsky
United Kingdom	Fisheries Radiobiological Laboratory, Lowestoft, Suffolk	J.W.R. Dutton
	British Nuclear Fuels Ltd., Windscale and Calder Works, Seascale, Cumberland	H. Howells
United States of America	Trapelo, Low Level Radiochemistry, Richmond, Calif.	W.J. Major
	Woods Hole Oceanographic Institution, Woods Hole, Mass.	V.T. Bowen
	US Environmental Protection Agency, Western Environmental Research Laboratory, Las Vegas, Nevada	M.W. Carter
	Eastern Environmental Radiation Laboratory, Montgomery, Alabama	D.G. Easterly
	US Atomic Energy Commission, Health and Safety Laboratory, New York, N.Y.	G.A. Welford

Country or Organization	Institution	Investigator
United States of America (cont.)	US Atomic Energy Commission, Health Services Laboratory, Analytical Chemistry Branch, Idaho Falls, Idaho	C.W. Sill
	Scripps Institution of Oceanography, La Jolla, Calif.	T.R. Folsom
IAEA	International Laboratory of Marine Radioactivity, Oceanographic Museum, Principality of Monaco	C.N. Murray

ANNEX II

ANALYTICAL METHODS USED BY DIFFERENT LABORATORIES FOR THE DETERMINATION OF PLUTONIUM IN SEAWATER AND SEAWEED SAMPLES

LABORATORY CODE No. 1

Seawater   Acid. with HNO_3; heat and add NH_4OH and then C_2H_5OH; filter ppt.; dissol. ppt. with $8\underline{M}$ HNO_3 + a few drops H_2O_2; anion exchange with DIAION S.R. 100 column; elute with $0.1\underline{M}$ Hl + $8\underline{M}$ HCl; evap. eluate to dryness; redissol. residue with H_2SO_4; adjust to pH 2 with NH_4OH; electrodeposition followed by alpha spectrometry.

Seaweed   Wet-ash with HNO_3 + $HClO_4$; follow seawater procedures.

LABORATORY CODE No. 2

Seawater   Spike ^{236}Pu; coppt. with $Fe(OH)_3$; fusion with KF; fuse cake with pyrosulphate; dissol. in HCl; coppt. with $BaSO_4$; dissol. in strong aluminium nitrate; Pu extract. into Aliquot-336; Pu stripped; electrodeposition followed by alpha spectrometry.

Seaweed   Wet ash with HNO_3 + H_2SO_4 + $HClO_4$ in presence of ^{236}Pu; evap. dryness; fusion with KF; follow seawater procedures.

LABORATORY CODE No. 3

Seawater   Spike ^{236}Pu; add H_2O_2; digest with $NaNO_3$; coppt. with $Fe(OH)_3$; centrifuge ppt.; dissol. ppt. with $12\underline{M}$ HCl; adjust volume to 60 ml with $6\underline{M}$ HCl; adjust molarity of HCl to $9\underline{M}$ with $12\underline{M}$ HCl; anion exchange with Dowex 1-X2 column; elute Fe with $7.2\underline{M}$ HNO_3; rinse column with $1.2\underline{M}$ HCl; elute Pu with 30% H_2O_2 + $1.2\underline{M}$ HCl; add 0.5 ml $18\underline{M}$ H_2SO_4 to eluate; evap. sol. until only H_2SO_4 remains; adjust pH to 2; electrodeposition followed by alpha spectrometry.

Seaweed   Ash sample + 20 ml $16\underline{M}$ HNO_3 + ^{236}Pu, digest; coppt. Pu on $Ca_3(PO_4)_2$ by using H_3PO_4; centrifuge and dissol. ppt. with $16\underline{M}$ HNO_3; evap. to dryness; add 30% H_2O_2 and $16\underline{M}$ HNO_3 alternately, allowing evap.; add 50 ml $6\underline{M}$ HCl to residue; boil and reduce vol. to 25 ml; add further 25 ml $6\underline{M}$ HCl; follow ion-exchange procedure as above.

LABORATORY CODE No. 5

Seawater   Weigh 1 kg of IAEA seawater into 1500 ml beaker. Add tracer ^{236}Pu. Add 100 ml $16\underline{N}$ HNO_3 and 50 ml $12\underline{N}$ HCl. Evap. sol. to small volume, salts begin to appear. Cool and add water to make $8\underline{N}$ with respect to HNO_3. Heat and add $NaNO_2$. Cool sample and add 5.0 ml Dowex 1 × 8 anion exchange resin. Filter off resin. Elute plutonium with 150 ml $0.4\underline{N}$ HNO_3 - $0.01\underline{N}$ HF. Evap. sol. to dryness. Add 15 ml. 1:1 HNO_3 at 0.25 ml 5% $NaNO_2$ sol. Add 2 ml resin. Filter off resin and wash with $12\underline{N}$ HCl and afterwards with 1:1 HNO_3. Elute plutonium with 100 ml of $0.4\underline{N}$ HNO_3 - $0.01\underline{N}$ HF. Convert to chloride, electrodeposition followed by alpha spectrometry.

Seaweed   Wet ash with 1.1 HNO_3 + HCl. Heat to decompose sample. Evap. small volume. Filter the undecomposed matter through Whatman 40. Ignite paper and residue. Treat residue with HF and HNO_3. evap. dryness. Dissolve residue with aluminium nitrate solution and combine with filtrate. Separate by Dowex ion-exchange. Electrodeposition followed by alpha spectrometry.

LABORATORY No. 6, No. 14*, No. 15* AND No. 17

Seawater Spike ^{236}Pu and add Fe-carrier; add NaHSO$_3$ and NH$_4$OH; recover hydroxide ppt. by centrifuge; dissol. ppt. with 16$\underline{M}$ HNO$_3$; dilute sol. with 8$\underline{M}$ HNO$_3$ to 100 ml; add 5 ml 30% H$_2$O$_2$; decompose H$_2$O$_2$ by heating; add NaNO$_2$; anion exchange with Dowex 1 × 8 column; elute Fe with 30 ml 8$\underline{M}$ HNO$_3$; elute Pu with conc. HCl + NH$_4$I; evap. to dryness; remove ammonium salt by evap. with conc. HNO$_3$ + conc. HCl; dissol. with conc. HCl; add NaNO$_2$; second anion exchange with Dowex 1 × 8; elute Pu with conc. HCl + NH$_4$I; evap. eluate; remove ammonium salts; electrodeposition followed by alpha spectrometry.

Seaweed Wet ash with H$_2$O and spike ^{236}Pu; add 200 ml 16$\underline{M}$ HNO$_3$ + 100 ml 12$\underline{M}$ HCl; digest for 2 h; add 300 ml 1$\underline{M}$ HNO$_3$ + 25 ml 30% H$_2$O$_2$; heat until decomp. of H$_2$O$_2$; filter through a glass fibre filter paper; rinse the residue with hot 1$\underline{M}$ HNO$_3$; digest residue with 16$\underline{M}$ HNO$_3$ + 12$\underline{M}$ HCl; combine the filtrates; evap. the filtrate until salts begin to form; dilute the sol. with 8$\underline{M}$ HNO$_3$; add 10 g NaNO$_2$; follow Pu purification procedure.

LABORATORY CODE No. 8

Seawater Spike ^{236}Pu; coppt. with Fe(OH)$_3$; dissolve ppt. with conc. HNO$_3$; evap. dryness; redissol. residue with conc. HNO$_3$; extract with T.O.A. acid; wash organic fraction; back extract with 0.1$\underline{M}$ HCl; wash aqueous phase with xylene, conc. HNO$_3$, conc. HClO$_4$; evap. dryness; redissol. residue; electrodeposition followed by alpha spectrometry.

Seaweed 30% H$_2$O$_2$ + sample, dissol. by heating; coppt. with Fe(OH)$_3$; follow as for seawater.

LABORATORY CODE No. 11

Seawater Spike ^{236}Pu; coppt. with Fe(OH)$_3$; dissol. ppt. and separate mixture of Fe and Pu with anion exchange; extract Pu with 1 phenyl-3-methyl-4 benzoyl-pyraxolone 5 in xylene; alpha spectrometry.

LABORATORY CODE No. 13

Seaweed Dry ashing; evap. to dryness with HF; fusion with KF + K$_2$S$_2$O$_7$; sep. of Pu on TIOA (tri-isooctyl-amine) ion exchanger; coppt. with LaF$_3$; alpha spectrometry.

LABORATORY CODE No. 16

Seaweed Coppt. with Fe-cupferrate; extract into CHCl$_3$; electrodeposition from an ammonium sulphate electrolyte; alpha spectrometry.

---

* Method applied for the seaweed sample only.

ANNEX III

REFERENCES QUOTED BY PARTICIPATING LABORATORIES FOR PLUTONIUM ANALYSIS IN ENVIRONMENTAL SAMPLES

HALLSTEIN, U., HOOGMA, A.H.M., KOOI, J., An improved method for the determination of trace quantities of plutonium in aqueous media-II. Interference of iron, calcium, uranium and chlorine; application to seawater and urine. Health Phys. 8 (1962) 45-49.

LAI, M.G., GOYA, H.A., Rep. USNRDL-TR-67-7 (1967).

MARKUSSEN, E.K., Radiochemical Procedures for the Determination of Plutonium in Environmental Samples, Rep. Risø-M 1242 (1970).

PILLAI, K.C., SMITH, R.C., FOLSOM, T.R., Plutonium in the marine environment, Nature (London) 203 (1964) 568-71.

SANDALLS, F.J., MORGAN, A., A New Procedure for the Routine Determination of Plutonium Alpha Activity in Urine, Using a Solid State Counter, Rep. AERE-R-4391 (1964).

TALVITIE, N.A., Radiochemical determination of plutonium in environmental and biological samples by ion exchange, Anal. Chem. 431 (1971) 1827-30.

TALVITIE, N.A., Electrodeposition of actinides for alpha spectrometric determination, Anal. Chem. 44 (1972) 280-83.

WONG, K.M., Radiochemical determination of plutonium in seawater, sediments and marine organisms, Anal. Chim. Acta 56 (1971) 355-64.

METHODS OF RUTHENIUM-106 ANALYSIS
IN MARINE ENVIRONMENTAL SAMPLES
A review

M. SHIOZAKI
Hydrographic Department,
Maritime Safety Agency

N. YAMAGATA, K. IWASHIMA
Department of Radiological Health,
Institute of Public Health,
Tokyo, Japan

Abstract

METHODS OF RUTHENIUM-106 ANALYSIS IN MARINE ENVIRONMENTAL SAMPLES: A REVIEW.
The analytical methods currently in use for the determination of ruthenium-106 in seawater, marine biota and sediments are critically reviewed. Pretreatment procedures for various matrices and purification steps of radioruthenium are discussed in detail. On the basis of the results of intercalibration exercises organized by the IAEA on radiochemical methods for analysis of fission products in seawater samples, the performance of various methods used by different laboratories are compared with each other. Based on the discussions in the text, recommended procedures for ruthenium-106 analyses by the authors are presented as an appendix.

1. INTRODUCTION

The analytical methods for the determination of stable and radioactive ruthenium in marine environmental samples were briefly reviewed by Chesselet in 1970 [1]. No new principle for analytical methods for radio-ruthenium has appeared since that time, although several improvements in analytical procedures have been proposed during this period. Thus, practically all methods reviewed in the present paper are those based on conventional methods described by Wyatt and Rickard [2]. Three reference methods proposed by the World Health Organization [3], the US Naval Radiological Defense Laboratory [4] and the Japanese Working Group on Chemistry of the Special Project Committee on the Release of Radioactive Materials into the Sea [5-7] were also included in the review.

Although the direct γ-spectrometry has been successfully applied to the determination of radioruthenium in marine organisms and sediments, especially in monitoring operations, the present review is mainly concerned with radiochemical procedures which are required for β-counting or γ-spectrometry, since the accuracy of direct γ-spectrometry depends mainly on types of instrumentation and modes of calibration, which are not specific for radioruthenium measurements.

2. REQUIRED DETECTION LIMITS AND SAMPLE SIZES

As has been given in the IAEA Technical Reports Series No.118, the required lower limit for the detection of ruthenium-106 in seawater is 0.2 pCi/l in a control context.

Iwashima [8] calculated on the basis of this limit the minimum sample size required for the determination of ruthenium-106 by using a low background gas flow counter, recommending the use of 2 litres of seawater, 1 g ash of marine organisms and 2 g dried sediments. Of course, a larger volume of seawater must be treated, when a γ-spectrometer or a β-counter without anti-coincidence shield is used.

3. DETERMINATION OF RUTHENIUM-106 IN SEAWATER

Ruthenium-106 occurring in seawater has been derived both from radioactive fall-out due to nuclear explosions and from radioactive waste effluents released from some nuclear installations. As the level of ruthenium-106 in seawater resulting from fall-out is, in general, very low, only a few measurements have been made. Where radioactive effluents have been released from nuclear fuel reprocessing plants [9], however, ruthenium-106 has become one of the most important radionuclides in a control context. Reflecting this importance of ruthenium-106 in monitoring operations, practically all analytical methods that have been published are for monitoring purposes.

Distillation or solvent extraction is usually used for the radiochemical separation of radioruthenium; in both cases the ruthenium-106 present originally in sample seawater and the ruthenium carrier have to be oxidized to ruthenium tetroxide in order to attain complete isotopic exchange between the radionuclide and carrier. This step is especially important for obtaining reliable radiochemical yields since the chemical forms of ruthenium-106 released are mixtures of many kinds of nitrosyl complexes [10].

A typical procedure for solvent extraction is that exploited by Loveridge and Thomas [11]. In this procedure the carrier and radioruthenium in a 1-litre sample are oxidized to perruthenate by adding potassium persulphate and potassium periodate in an extremely alkaline medium and successive heating to 90-95°C; after adjusting the pH to 4-6, ruthenium is extracted by carbon tetrachloride and then precipitated as dioxide; after mounting and counting, the chemical yield of ruthenium is determined by spectrophotometric measurement of a solution prepared from the source. The lower limit of detection of this procedure is 5 pCi ruthenium-106. This method has been adopted as the reference method by WHO [3].

Kiba et al. [12] applied the above-mentioned method to 3 litres of surface seawater with several modifications. They used 60 min heating time and 300-600 ml of carbon tetrachloride to increase the extractability of ruthenium tetroxide. The extracted RuO_4 is back-extracted with a potassium hydroxide solution containing potassium periodate and then ruthenium is reduced by adding ethanol and successive warming. The precipitate is filtered with an MF filter and mounted. After measuring the activity, the determination of the chemical yield is carried out in the same way as in Loveridge's method [11].

Iwashima [8] recommended the following procedures on the basis of the critical review of proposed methods and his own experiments. The oxidation and solvent extraction procedures are similar to those of Loveridge's method [11]; the extracted ruthenium is back-extracted with a sodium hydroxide solution containing reducing agent; hydrous ruthenium oxides are precipitated by adding alcohol to back-extracts and warming and the precipitates formed are filtered; after drying, the radioactivity is measured; the chemical yield is determined by spectrophotometric measurement using a known portion of back-extracted solution.

The method adopted by the US Naval Radiological Defense Laboratory [4] utilizes the property of ruthenium that dissolves in oxidizing alkaline media. In this method radioruthenium in seawater and added carrier are converted to a mixture of ruthenate and perruthenate, which are soluble in alkaline media and can be separated from many other fission products usually scavenged by coprecipitation with ferric and zirconium hydroxides [13]. This scavenging procedure is repeated with zirconium hydroxide. The ruthenium is isolated as hydrated ruthenium oxides. The oxides are dissolved in hydrochloric acid and then the ruthenium is reduced to metal with magnesium metal for weighing and counting.

Kiba et al. [14, 15] used an iron particle (80-120 mesh) column to collect the radioruthenium (nitrosylruthenium trinitrate) from 50 ml of artificial seawater. The column adsorbed 86 and 97% of the added radio-ruthenium at flow rates of 1 ml/min and 0.3 ml/min, respectively. The adsorbed ruthenium is eluated with a mixture of sodium hydroxide solution and antiformin and then extracted with pyridine.

Watari et al. [16, 17] used copper sulphide-ion exchange resin and ferric hydroxide-cation exchange resin for the effective preconcentration of radioruthenium from seawater.

There are only a few examples of the determination of ruthenium-106 in seawater for research purposes.

Yamagata and Iwashima [18] determined radioruthenium in the harbours of Sasebo and Yokosuka in Japan to estimate the concentration factors of marine organisms. An outline of their method is as follows: radioruthenium in a large volume of seawater was coprecipitated with ferric hydroxide and then ferric ions were eliminated by extracting into isopropyl ether. Ruthenium was purified by distillation and solvent extraction procedures and chemical yield was determined by spectrophotometric methods after β-counting.

Shiozaki [19] also carried out the determinations of ruthenium-106 in seawater for research purposes. The radioruthenium in seawater and added carrier are oxidized to perruthenate by adding antiformin at a pH at which a part of magnesium hydroxide slowly precipitates; perruthenate is reduced by sodium bisulphite and ethyl alcohol and precipitated from 20-40 litres of seawater by coprecipitation with magnesium hydroxide, and then separated from other interfering nuclides by oxidation to perruthenate in alkaline media and successive solvent extraction of ruthenium tetroxide with carbon tetrachloride. Back extraction, chemical form for counting and yield determination procedures are as described by Iwashima [8].

Another example of an analytical method for ruthenium-106 in seawater is that proposed by Shiozaki, Seto and the Chemical Working Group of the Special Project Committee on the Release of Radioactive Materials into the Sea [5]. This method is a sequential analytical method of cerium-144

TABLE I. COMPARISON OF THE RESULTS OF RUTHENIUM-106
MEASUREMENTS ON SEAWATER WITH AND WITHOUT OXIDATIVE-
REDUCTION PRETREATMENT

	^{106}Ru (pCi/l)	
	With pretreatment	Without pretreatment
Kuroshio	0.66 ± 0.03	0.33 ± 0.02
	0.17 ± 0.01	0.14 ± 0.01
Japan Sea	0.80 ± 0.03	0.69 ± 0.03
	1.20 ± 0.04	0.96 ± 0.04

and ruthenium-106 in seawater and is designed to allow the determination
of fall-out level of ruthenium-106 effectively. The principle of this method
is the same as that used by Shiozaki [19].

Although direct γ-spectrometry is the most desirable method for routine
environmental monitoring, it is restricted to samples of high radioactivity.
In the case of comparatively high levels of radioruthenium γ-spectrometry
has been applied after evaporation of sample water to dryness [20-22].
At low levels, however, preconcentration of radioruthenium by coprecipitation
or adsorption [21] was necessary. Sulphide precipitation has been used as
the most promising technique for the preliminary concentration of ruthenium
for γ-spectrometric determination. Sodd et al. [23] applied homogeneous
sulphide precipitation using thioacetamide for the measurement of gross
radioactivity in seawater. Chakravarti et al. [24] used sulphide precipitation
caused by bubbling through hydrogen sulphide at pH 11 for the determination
of a number of radionuclides including ruthenium. Yamagata and
Iwashima [25] used cobalt sulphide, which was precipitated homogeneously
with thioacetamide at pH 9.5 for the preconcentration of radioruthenium
from 5-10 litres of seawater for γ-spectrometry. The homogeneous sulphide
precipitation technique with thioacetamide was originally developed by
Swift and Butler [26].

4.   THE DETERMINATION OF RUTHENIUM-106 IN BIOLOGICAL
     MATERIALS

Almost all methods for the radiochemical determination of ruthenium-106
in marine organisms use ashing and fusion as pretreatment steps and then
solvent extraction as purification procedure.

Tsuruga [27-29] ashed biological materials at 450°C. The ashed sample,
to which the carrier and potassium hydroxide, carbonate and nitrate were
added, was fused according to the method described by Meritt [30]. The
melt was leached with hot water and ruthenium was precipitated as sulphide.
The precipitate was dissolved in 9N nitric acid and then evaporated to
dryness. The residue was dissolved in hydrochloric acid and ruthenium
was precipitated as metal by adding magnesium metal. The ruthenium metal
was filtered, dried, weighed and counted with a low-background gas-flow
counter.

Iwashima [8] adopted solvent extraction of ruthenium tetroxide with carbon tetrachloride after the alkali fusion of the ash. Acetic acid-sodium acetate buffer solution was used to hold the pH of the aqueous layer at about 4, at which the highest extractability of ruthenium was backextracted to a 3$\underline{M}$ sodium hydroxide solution containing sodium bisulphite. The succeeding procedures were similar to those for seawater described by the same author. This method has also been adopted as the reference method by the Chemistry Working Group of the Special Project Committee on the Release of Radioactive Materials into the Sea [7]. Another analytical method, which was proposed by WHO, also adopts fusion of the ash and successive solvent extraction of ruthenium [3].

5. THE DETERMINATION OF RUTHENIUM-106 IN MARINE SEDIMENTS

Distillation of ruthenium tetroxides [31-33] and solvent extraction have been applied to the radiochemical analysis of radioruthenium in sediments. Although various workers [34,35] have used direct γ-spectrometry for the determination of radioruthenium in sediments, the counting conditions have not been described in detail.

In a control context about 1-2 g of dried sediments are enough to determine ruthenium-106 by using a low-background gas-flow counter after chemical purification [8].

The most reliable method for the determination of ruthenium-106 in sediments seems to be the fusion method. Iwashima [8] recommended the fusion and solvent extraction method on the basis of a critical review of the previously proposed methods. In this method organic materials present in sediments are ashed at 400-500°C and then fused with a mixture of potassium hydroxide and potassium nitrate. The ruthenium in the melt is leached with water containing sodium hypochlorite solution and then extracted into carbon tetrachloride.

The method for the determination of radioruthenium in sediments proposed by WHO [3] also utilizes fusion and solvent extraction techniques. The proposed procedures are similar to those used for the analysis of ashed marine organisms.

Direct distillation techniques of ruthenium tetroxide from sediments have been adopted by several investigators for the determination of radio-ruthenium in sediments and soil [6, 12, 36-38]. Kahn [36] used dilute nitric acid, dilute or concentrated hydrochloric acid in order to leach ruthenium-106 adsorbed on soil but the results were unsatisfactory. The satisfactory removal of ruthenium from soil was attained by adding ruthenium carrier to a leaching solution (9$\underline{M}$ H$_2$SO$_4$), oxidizing ruthenium with potassium permanganate, and distilling the ruthenium tetroxide into a sodium hydroxide solution.

Shiozaki et al. [37] modified Kahn's method and applied it to marine sediments. Organic matter in the sediment was decomposed by heating the sample with conc. sulphuric and nitric acids, and at the same time chloride, which consumes permanganate, was eliminated as hydrochloric acid gas. Ruthenium-106 in the sample and added carrier are distilled as ruthenium tetroxide by oxidizing with potassium permanganate. The ruthenium tetroxide generated was absorbed into a 3$\underline{M}$ sodium hydroxide solution and then reduced by adding ethanol. The precipitates were filtered,

dried and β-counted. The chemical yield was determined spectrophotometrically, using a portion of the solution of absorbed ruthenium tetroxide.

The direct distillation method was also reported by Kiba et al. [12].
Osmium and ruthenium in marine sediments were sequentially and
separately distilled by two different oxidizing agents at the presence of
strong phosphoric acid; ceric sulphate for osmium and potassium dichromate
for ruthenium were used as the oxidizing agents. The distilled ruthenium
was adsorbed in a mixture of 6 M hydrochloric acid and ethylalcohol.
Ruthenium was finally precipitated as metal, filtered, dried for weighing
and β-counted. The recovery of ruthenium was constantly quantitative.

Another distillation procedure was devised by Nishiya et al. [6]. The
radioruthenium in marine sands is leached with nitric acid by boiling and
ruthenium in the leaching solution is distilled by boiling with sulphuric
acid and potassium permanganate. The distilled ruthenium tetroxide is
absorbed in a sodium hydroxide solution. A portion of the solution is used
for chemical yield determinations by the spectrophotometric method and
the other portion is used for β-counting after precipitation, filtration and
weighing. This method was adopted as the reference method by the Japanese
Chemistry Working Group of the Special Committee.

Kautsky [39] used aqua regia to leach ruthenium-106 from solid samples.
Ruthenium is precipitated as sulphide, dissolved in hydrochloric acid and
then distilled by using perchloric acid as oxidant. The ruthenate formed
is reduced to metal with zinc metal in a hydrochloric acid solution and
ruthenium metal is separated for weighing and radioactivity measurement.

6.   DISCUSSION

6.1. Adsorption of radioruthenium to the container wall

According of Shiozaki [19], 95% of added ruthenium-106 was recovered
when the seawater sample was acidified to pH 1.5 and stored for 2 days in
a polyethylene bottle, while only 56% of the ruthenium-106 was recovered
at pH 6.

Iwashima [3] carried out experiments on the adsorption of ruthenium-106
in several chemical forms, such as chlorocomplexes, nitrosyl chlorocomplexes and nitrocomplexes, to the wall of polyethylene bottles at pH
of 1.5, 2.0 and 8.2. The results showed that only 89.2 and 74.3% nitrosyl
chlorocomplexes of ruthenium-106 were recovered respectively after
2 and 12 days of storage at pH 8.2, while added ruthenium-106 was almost
quantitatively recovered in other cases. These results indicate that the
pH of the seawater should be adjusted to 0.5-2.0 for storage.

6.2. Pretreatment

6.2.1. Seawater

As many physico-chemical states of radioruthenium may be present
in seawater [40-46], oxidation and reduction pretreatments must be carried
out to ensure complete isotopic exchange between radioruthenium in seawater and the added carrier.

Shiozaki [19] observed that ruthenium-106 determined in seawater was increased by applying oxidation with antiformin and reduction with sodium bisulphite and ethanol to water samples before the concentration of ruthenium. The results are shown in Table I. These differences may be due to the presence of organic chelate compounds (though their presence was not confirmed [42,45]) or inorganic nitrosyl ruthenium nitrocomplexes. To eliminate these effects, it is necessary to oxidize any chemical form of ruthenium to perruthenate and decompose the organic materials. These requirements are fulfilled by the method of oxidation devised by Loveridge and Thomas [11]. To facilitate the oxidation, the reaction system is heated but never boiled in order to avoid the loss of ruthenium. This oxidation procedure was adopted by WHO [3] and by Iwashima [8]. Since in this procedure about 14 nCi of potassium-40 are added to a sample as potassium persulphate, hydroxide and periodate, washing of the organic layer must be made carefully to eliminate the contamination by potassium-40. Presumably, to avoid the contamination by potassium-40, Lai and Goya [4] used sodium hydroxide with boiling for 15 min. According to Iwashima [8], however, there may be a loss of ruthenium through boiling.

Shiozaki [19] used sodium hypochlorite as oxidizing agent and confirmed spectrophotometrically that the ruthenium was oxidized to perruthenate [33]. Heating was avoided to prevent the evaporation of ruthenium tetroxide, which may be present [17]. In this oxidizing condition organic matter like simple amino acids may be decomposed, while oxidation may not be perfect for other organic substances.

Iron particle columns [14], anion exchange columns saturated with copper sulphide [16], and cation exchange columns saturated with ferric hydroxide [17] are also used for the concentration of ruthenium with other nuclides in seawater. Although it was confirmed that these methods are effective for the concentration of nitrosyl ruthenium nitrocomplexes, there are only a few examples of the application to environmental samples.

Two methods of preconcentration for γ-spectrometry may be cited: evaporation and coprecipitation. The evaporation method is time-consuming and limited to relatively high-level samples because of the large quantities of deposited salts. Coprecipitation is a rapid procedure, though there is a problem of whether or not the nitrosyl ruthenium nitrocomplexes are easily coprecipitated. Yamagata and Iwashima [25] confirmed the quantitative coprecipitation of several kinds of nitrosyl ruthenium with uniform precipitation of cobalt sulphide at pH 9.5. Similar results were also obtained by Watari et al. [48]. By using this procedure 5-10 litres of seawater can be treated within a few hours.

6.2.2. Biological materials

Strohal and Jelisavcic [49] made an ashing experiment on a mollusc that was allowed to take up radioruthenium from seawater. They recovered only 89 and 70% of the ruthenium at 110 and 450°C, respectively. Riley [50] also made similar experiments using seaweed at 450-500°C, observing that the loss of radioruthenium during the ashing procedure did not exceed 5%.

Iwashima [8] conducted tracer experiments to examine the results obtained by Strohal and Jelisavcic [49] under similar conditions. He observed that 96% or ruthenium-106 incorporated in mussels remained on drying at 110°C and 92% on ashing at 450-500°C for 18 hours. Experiments

on the loss of ruthenium-106 during the ashing process were also reported
by Nakamura et al. [51]. Bivalves and green algae were contaminated by
rearing the organisms in seawater containing various radionuclides and
they were ashed at 450, 550 or 800°C for 12, 24 or 48 hours, respectively.
The results showed that the degree of loss of radioruthenium is influenced
by ashing condition and species of organisms. They recommended ashing
at 550°C for 12 hours for ruthenium in bivalves and at 450°C for 24 hours
for that in green algae. The ashing methods recommended by WHO [3]
or adopted by Tsuruga [29] are similar to the above-mentioned method.

6.2.3. Sediments

Although complete isotopic exchange between radioruthenium in sediments and added ruthenium carrier is assured by the fusion method, it
has not been proved that this is the case for the distillation method.

6.3. Purification and other procedures

The solvent extraction method of RuO_4 with carbon tetrachloride is
preferable to the distillation method since the time required for the
extraction is shorter and decontamination from other nuclides can be
obtained by the extraction [52].

In the method adopted by WHO [3] the extracted ruthenium is directly
precipitated in the organic layer and the chemical yield is determined after
counting, while in the method of Iwashima [8] ruthenium is back-extracted
in a 3M sodium hydroxide solution containing sodium bisulphite. Two
methods of chemical yield determination have been reported; one is the
spectrophotometric method of Ru(IV) [53], and the other is the weighing
of ruthenium metal [4, 29, 38]. In the spectrophotometric method counting
for radioactivity measurement must be made quantitatively. When weighing
ruthenium metal it is preferable to use more than 30 mg ruthenium carrier
and not to use filter paper for accurate yield determination.

When the radioruthenium in the sample and the added carrier are
oxidized to perruthenate there seems to be a difference between the oxidation
rate of the radioruthenium and that of the carrier owing to the isotope
effect [52]. The oxidation time, therefore, must be prolonged sufficiently
to offset the isotope effect. Iwashima [8] studied the equilibrium between
radioruthenium and ruthenium carrier by comparing the spectrophotometrically measured yield with that radiometrically obtained. Iwashima [8]
also determined the decontamination factors for both fission and activation
products in seawater and biological samples and also those for the natural
radionuclides present in marine sediments. The results show that the
decontamination of radioruthenium by the extraction method from the nuclides
tested was satisfactory.

When 10-20 g of sediments have to be treated the direct distillation
method is more practical than the fusion method.

Nagaya and Saiki [34] studies the leaching conditions of several fission
products including ruthenium-106 in 100 g of mud collected from Tokyo Bay.
They concluded that 96% of ruthenium-106 (fall-out) was leached by 6 hours
digestion with hot hydrochloric acid (1:1), but the succeeding ruthenium
purification procedures were not described.

TABLE II. RESULTS OF RUTHENIUM-106 MEASUREMENTS REPORTED BY VARIOUS LABORATORIES

Method		^{106}Ru in seawater (pCi/kg)	
Counting	Chemical	SW-I-1	SW-I-2
Gamma spectrometry	Direct	32	59 ± 33
		14	54.7 ± 14.8
		1.6 ± 0.2	19.5 ± 0.6
	Evaporation residue	2.7 ± 0.6	41.1 ± 0.6
		14	62 ± 10
		N.D.	45.6 ± 2.8
		—	34.9 ± 4.5
		5.7 ± 2.4	47.5 ± 4.9
	Ion-exchange column (Chelex-100)	0.50 ± 0.68	3.37 ± 0.39
	MnO_2 ads.	28.5 ± 3.5	38.5 ± 3.2
		4	N.D.
	CUFC ppt.	N.D.	N.D.
	Co-sulphide ppt.	2.4 ± 0.4	39.9 ± 1.3
		3.0 ± 0.2	31.7 ± 1.2
	$Fe(OH)_3$ ppt.	9.6 ± 3.7	29.9 ± 2.6
		0.7	0.3
Beta counting	$Fe(OH)_3$ ppt. Ru-sulphide ppt. CCl_4 extr. of RuO_4, Ru	1.3 ± 0.2	13.7 ± 0.6
	$Fe(OH)_3$ ppt. Ru-sulphide ppt. dist. of RuO_4, Ru	1.3 ± 0.1	12.2 ± 0.7
	$Fe(OH)_3$ ppt. dist. of RuO_4, RuO_2 ppt.	1.7 ± 0.1	38.0 ± 1.2
	Mixed hydroxide-oxide ppt., CCl_4 extr. of RuO_4, RuO_2	3.35 ± 0.30	43.7 ± 0.5
	Mixed hydroxide-oxide ppt., dist. of RuO_4, Ru	3.5 ± 0.3	41.7 ± 1.2
		1.44 ± 0.15	7.48 ± 0.24
	RuO_2 ppt., oxid., CCl_4 extr. of RuO_4, RuO_2 ppt.,	3.4 ± 0.5	34.2 ± 2.2
	Oxid., CCl_4 extr. of RuO_4, RuO_2 ppt., Ru	2.4 ± 0.4	39.9 ± 1.3
		—	38.6 ± 1.3

6.4. Comparison of various methods

To obtain some idea of the performance of various analytical methods
used currently by different laboratories, it is useful to compare the results
of the intercalibration exercise organized by the Monaco Laboratory on
radiochemical methods for analysis of fission products in seawater samples.
The results of ruthenium-106 measurements reported by different laboratories
are classified according to the methods used and shown in Table II.

The following observations have been made on the results presented in
Table II:

(i)   Generally speaking, the levels of radioactivity in the distributed
      samples are too low for direct γ-spectrometry;

(ii) For γ-spectrometry of the higher level sample evaporation to dry-
      ness is sufficient to measure ruthenium-106 activity. The results
      obtained agree well with each other within the estimated errors;

(iii) The adsorption methods to ion exchange columns or CUFC give
      significantly low results;

(iv) Sulphide precipitation as the preconcentration method for
      γ-spectrometry seems to be satisfactory and gives good results;

(v)   The results of β-counting after various chemical purification steps
      do not agree with each other. Perhaps these scattered results are
      related to the different effectiveness of sample pretreatment rather
      than differences in purification procedures;

(vi) Although it seems that relatively low results are obtained when
      ruthenium metal is used as final chemical form for weighing and
      counting, this may not be the significant cause of the scattered results;

(vii) More detailed description of the procedures must be available before
      the cause of the disparity in results can be determined.

7. CONCLUSIONS

7.1. Gamma spectrometry

7.1.1. Seawater

Samples containing more than 50-100 pCi radioruthenium per sample
can be analysed directly by γ-spectrometry after evaporation. For samples
of larger volumes (5-10 litres) the preconcentration procedure of radioruthenium with cobalt sulphide is recommended.

7.1.2. Marine organisms

The direct γ-spectrometry of radioruthenium is, in general, more
easily applicable to samples of marine organisms than to those of seawater because of their higher activity levels. For low radioactivity samples
ashing at 450-500°C for 8-24 hours before γ-spectrometry is recommended.

7.1.3. Marine sediments

The direct γ-spectrometry of radioruthenium in marine sediments is
not an easy task, unless samples contain sufficient radioruthenium in

relation to the radionuclides naturally occurring in sediment samples. These natural radionuclides make the estimation of ruthenium-106 in the photopeak area of 0.62 MeV difficult, especially when a NaI(Tl) detector is used.

7.2. Radiochemical procedures

7.2.1. Seawater

The oxidation procedure for radioruthenium with its carrier added to seawater samples and the successive solvent extraction procedure, both described in the Appendix, seem to be preferable to other procedures used by various workers for seawater in view of the completeness of the isotopic exchange between radioruthenium and its carrier, as well as the ease of the analytical operations. For chemical yield determinations the gravimetric method of ruthenium metal is recommended for its good reproducibility.

7.2.2. Marine organisms

The fusion procedure of ash with potassium nitrate and potassium hydroxide, successive solvent extraction procedures and gravimetric chemical yield determination, described in the Appendix, are recommended for use with marine organisms.

7.2.3. Marine sediments

Although a similar method to that for marine organisms is recommended for marine sediments, a direct distillation method of ruthenium as RuO_4 is also attractive in view of its simple and rapid performance when more than 10 g of sediments have to be analysed.

APPENDIX

RADIOCHEMICAL ANALYSIS OF RUTHENIUM
(Recommended procedures by the Authors)

SHELLFISH AND SEAWEED

1.   Remove adhered water and wash quickly with distilled water. Blot with filter paper to remove adhered water and weigh to obtain fresh weight.

2.   Dry at 110°C and put into a porcelain dish. Heat with a burner to char until no further gas evolves (about 400°C). Care should be taken not to burn.

3.   Put the dish into an electric furnace and ash at 450-500°C. After cooling weigh to obtain ash content.

4.   Weigh 1 g ash into a nickel crucible and add 2 g each of potassium nitrate and potassium hydroxide and 2 ml of ruthenium carrier solution (10 mg Ru). Mix well, dry by an infra-red lamp and then fuse for one hour at 500°C in a muffle furnace.

5.   After cooling, add about 20 ml of water in several portions to the crucible and heat. Transfer the contents of the crucible into a 50 ml centrifugal tube.

6.   Add 5 ml of 10% sodium hypochlorite solution, mix and stand for half an hour.

7.   Centrifuge for 2 min at 3000 rev/min and transfer the supernatant to a 100 ml separatory funnel (A).

8.   Add one drop each of $3M$ sodium hydroxide and 10% sodium hypochlorite solution and 10 ml of water to the residue in the centrifugal tube, mix well and then centrifuge for 2 min at 3000 rev/min. Transfer the supernatant to the separatory funnel (A).

9.   Add 30 ml of carbon tetrachloride, 5 ml of $6M$ hydrochloric acid and 10 ml of $3M$ acetic acid-sodium acetate buffer solution to the separatory funnel (A). Shake the funnel for one minute and then transfer the organic phase to another 100 ml separatory funnel (B).

10.   Add 10 ml carbon tetrachloride to the aqueous phase in the funnel (A), shake for one minute and then combine the organic phase with that in the funnel (B).

11.   Add 20 ml of water containing one drop of 10% sodium hypochlorite solution to the organic phase in the funnel (B) and shake for one minute. Transfer the organic phase to another separatory funnel (C). Discard the aqeuous phase.

12.   Add 20 ml of $3M$ sodium hydroxide containing 3 drops of $1M$ sodium hydrogen sulphite and shake for one minute. If the colour remains in the organic phase, add another drop of $1M$ sodium hydrogen sulphite and shake for another minute.

13.   Transfer the aqueous phase to a 50 ml glass centrifuge tube and add 2 ml of ethyl alcohol. Digest on a water bath until the precipitate coagulates.

14.   Centrifuge the precipitate and discard supernatant. Dissolve the precipitate in 2 ml of HCl by warming and dilute to 20 ml with water.

15.   Add powdered magnesium metal in small portions until the solution is colourless.

16.   Add HCl drop-wise until all the excess magnesium metal dissolves. Boil, cool and centrifuge.

17.   Wash the precipitate three times with hot water and once with ethyl alcohol.

18.   Transfer the precipitate with ethyl alcohol to a tared planchet (C g), dry, weigh (D g), and count ($N \pm \Delta N$ counts/min).

BOTTOM SEDIMENTS

1.   Dry sediment sample without heating. Grind the dried sediment and pass through 2 mm sieve.

2.   Take about 2 g sample in nickel crucible, dry at 110°C and weigh after cooling.

3.   Add 2 ml ruthenium carrier solution (10 mg Ru), mix well, dry by an infra-red lamp and then ignite at 450-500°C for one hour in an electric furnace.

4. After cooling, add 4 g each of potassium hydroxide and potassium nitrate and mix well. Fuse for one hour at 500°C in a muffle furnace.

5. After cooling, add about 30 ml of water in several portions to the crucible and heat. Transfer the contents of the crucible into a 50 ml centrifugal tube.

6. Add 8 ml of 10% sodium hypochlorite solution, mix and stand for half an hour.

7. Centrifuge for 5 min at 3000 rev/min and transfer the supernatant to a 150 ml separatory funnel (A).

8. Add one drop each of 3M sodium hydroxide and 10% sodium hypochlorite solution and 10 ml of water to the residue in the centrifugal tube, mix well and then centrifuge for 2 min at 3000 rev/min. Transfer the supernatant to the separatory funnel (A).

9. Add 30 ml of carbon tetrachloride, 10 ml of 6M hydrochloric acid and 10 ml of 3M acetic acid-sodium acetate buffer solution to the separatory funnel (A). Shake the funnel for one minute and then transfer the organic phase to another 100 ml separatory funnel (B).

10. Add 15 ml carbon tetrachloride to separatory funnel (A), shake for one minute and then combine the organic phase with that in the funnel (B). Follow the procedures from step (11) described under Shellfish and Seaweed.

SEAWATER

1. After collection of 2 l seawater, filter immediately through a millipore filter of 0.45 μm pore size. Take the filtrate in a polyethylene bottle and add 2 ml concentrated hydrochloric acid to make the pH approximately 1.5, then transfer to laboratory.

2. Pour seawater into a 3-litre beaker, wash the polyethylene bottle with 150 ml of 3M hydrochloric acid and combine the washing with the seawater.

3. Add 2 ml of ruthenium carrier solution (10 mg Ru) and heat the solution almost to boiling with occasional stirring.

4. After cooling, add 65 g potassium hydroxide, 20 g potassium peroxodisulphate and 10 g potassium periodate.

5. Cover the beaker with a watch glass and heat the solution at 90-95°C (do not boil) for 2 hours. Wash the watch glass with 10% sodium hypochlorite solution and combine the wash with the solution.

6. Cool the solution with ice-water and transfer to a 3-litre separatory funnel (A). Add 200 ml carbontetrachloride, adjust pH to 4-5 by adding 6M nitric acid and shake for 5 min.

7. Transfer the organic phase to another separatory funnel (B). Add 50 ml carbontetrachloride to the aqueous phase in the funnel (A), shake for one minute and then combine the organic phase with that in the funnel (B). Discard the aqueous phase.

8. Add 150 ml of water containing 15 drops of 10% sodium hypochlorite solution to the organic phase in the funnel (B) and shake one minute. Transfer the organic phase to another separatory funnel (C). Discard the aqueous phase.

9. Add 25 ml of 3M sodium hydroxide containing several drops of 1M sodium hydrogen sulphite and shake for one minute. Transfer the organic phase to another separatory funnel (D) and take the aqueous phase in a 100 ml centrifugal tube.

10. Repeat the first half of step (9) on the organic phase in the funnel (D). Discard the organic phase and combine the aqueous phase with the solution in the centrifugal tube. Add 2 ml of ethyl alcohol and digest on a water bath until the precipitate coagulates. Follow the procedures from step (14) described under Shellfish and Seaweed.

STANDARDIZATION

1. Take 2 ml of ruthenium carrier solution (10 mg Ru) in a 50 ml centrifugal tube and add 1 ml of [106]Ru standard solution (a dpm) and about 10 ml of water.

2.   Add 10M sodium hydroxide until black precipitate forms and then 10 drops excessively.

3.   Add 10% sodium hypochlorite solution drop-wise to dissolve the precipitate, stand for half an hour, add 2 ml ethanol and then heat for half an hour on a water bath to precipitate hydrous ruthenium oxides.

4.   Measure the radioactivity (b counts/min) as described in steps (14) – (18) under Shellfish and Seaweed.

5.   Calculate the counting efficiency (E%) by the equation:

$$E = b/a \times 100$$

CALCULATION

1.   Calculate the activity of ^{106}Ru (pCi/g or 1 litre) in the sample by the equation:

$$^{106}Ru = (N \pm \triangle N) \, e \, \gamma \, t \, \frac{1}{M} \, \frac{100}{E} \, \frac{100}{Y} \, \frac{1}{2.22}$$

    γ: decay constant for ruthenium-106
    t: year between sampling and measurement
    M: Sample weight (g) or volume (l).
    E: Counting efficiency (%)
    Y: Chemical yield (%)

RAPID METHOD FOR SEAWATER BY γ-SPECTROMETRY

1.   After collection of 5 litres of seawater, filter immediately through a millipore filter of 0. 45 μm pore size. Take the filtrate in a polyethylene bottle and add 5 ml conc. HCl to make the pH approximately 1. 5, then transport to laboratory.

2.   Pour seawater into a 5-litre beaker, wash the polyethylene bottle with about 400 ml of 3M hydro-chloric acid and combine the washing with seawater.

3.   Add 5 ml of cobalt carrier solution ($CoCl_2$, 10 mg Co/ml), heat to boil and then adjust the pH to about 9. 7 by addition of 7M ammonia water.

4.   Heat at 90-95°C for 2 hours on a hot plate after the addition of 30 ml of 4% (w/v) thioacetamide solution.

5.   Filter the precipitate through a millipore filter of 0. 45 μm pore size and 47 mm in diameter by suction, wash with 10 ml of water containing one drop of 1M ammonia water, dry and measure the peak activity at 0. 62 MeV.

REFERENCES

[1] CHESSELET, R., "Review of the techniques of measuring stable and radioactive Ru, Mn, Cr, Fe and Zr-Nb in a marine environment", Reference Methods for Marine Radioactivity Studies, Technical Reports Series No. 118, IAEA, Vienna (1970) 275-84.

[2] WYATT, E. I., RICKARD, R. R., The Radio-chemistry of Ruthenium, Rep. NAS-NS 3029 (1961).

[3] WORLD HEALTH ORGANIZATION, Methods of Radiochemical Analysis, Part II, Analytical Methods, 4, Ruthenium, WHO, Geneva (1966) 84-93.

[4] LAI, M. G., GOYA, M. A., A Compendium of Radiochemical Procedures for the Determination of Selected Fission Products in Sea Water", US Naval Radiological Defense Laboratory (1965) 26-29.

[5] SHIOZAKI, M., SETO, Y., CHEMISTRY WORKING GROUP of the Special Project Committee on the Release of Radioactive Materials into the Sea, Sequential Analytical Method of Radiocerium and Radioruthenium in Sea Water, Japan Atomic Energy Safety Res. Assoc. (1970).

[6] NISHIYA, T., YAMATO, A., IWASHIMA, K., YAMAGATA, N., CHEMISTRY WORKING GROUP of the Special Project Committee on the Release of Radioactive Materials into the Sea, Analytical Method of Radioruthenium in Sea-bed Sand, Japan Atomic Energy Safety Res. Assoc. (1970).

[7] IWASHIMA, K., KOYAMA, K., CHEMISTRY WORKING GROUP of the Special Project Committee on the Release of Radioactive Materials into the Sea, Analytical Method of Radioruthenium in Marine Organisms, Japan Atomic Energy Safety Res. Assoc. (1969).

[8] IWASHIMA, K., Analytical methods for ruthenium-106 in marine samples, J. Radiat. Res. 13 (1972) 127-48.

[9] HOWELLS, H., "Discharges of low-activity, radioactive effluent from the Windscale works into the Irish sea", Disposal of Radioactive Waste into Seas, Oceans and Surface Waters (Proc. Symp. Vienna, 1966), IAEA, Vienna (1966) 769-85.

[10] BROWN, P. G. M., FLETCHER, J. M., HARDY, C. J., DENNEDY, J., SCARGILL, D., WAIN, A. G., WOODHEAD, J. L., "The significance of certain complexes of ruthenium, niobium, zirconium, and uranium in plant processes", Peaceful Uses of Atomic Energy (Proc. Conf. Geneva, 1958) 17, UN, New York (1958) 118-29.

[11] LOVERIDGE, B. A., THOMAS, A. M., The Determination of Radioruthenium in Effluent and Sea Water, Rep. AERE C/R 2828 (1959).

[12] KIBA, T., TERADA, K., KIBA, T., "The determination of radioruthenium in sea water and sediments", Abstract 3rd Meeting Atomic Energy Safety Research, Japan Atomic Safety Res. Assoc. (1970) 55.

[13] RICKARD, R. R., WYATT, E. I., Radiochemical determination of fission ruthenium in aqueous solutions. A nondistillation technique, Anal. Chem. 31 1 (1959) 50.

[14] KIBA, T., UCHISHIMA, H., SUGIOKA, Y., Abstract 14th Annual Assembly Chem. Soc. Japan (1961).

[15] KIBA, T., MIURA, A., SUGIOKA, Y., The separation and determination of ruthenium in fission products by liquid-liquid extraction with pyridine, Bull. Chem. Soc. Japan 36 6 (1963) 663-69.

[16] WATARI, K., TSUBOTA, H., KOYANAKI, T., IZAWA, M., Concentration of radionuclides in sea water by metal sulfide-ion-exchange resins, J. At. Energ. Soc. Japan 8 4 (1966) 182-85.

[17] WATARI, K., TSUBOTA, H., KOYANAGI, T., IZAWA, M., Concentration of radionuclides in sea water by ferric hydroxide cation exchange resin, J. At. Energ. Soc. Japan 8 3 (1966) 130-33.

[18] YAMAGATA, N., IWASHIMA, K., Radioactive contamination in the harbours of Sasebo and Yokosuka, Bull. Inst. Publ. Health 14 4 (1965) 183-93.

[19] SHIOZAKI, M., Ruthenium-106 in the adjacent Sea of Japan, Rep. Hydrographic Res. 1 (1966) 33-45.

[20] DUTTON, J. W. R., Gamma Spectrometric Analysis of Environmental Materials, Fisheries Radiological Laboratory Rep. FRL 4 (1969).

[21] PICKERING, R. J., CARRIGAN, P. H., TAMURA, T., ABEE, H. H., BEVERAGE, J. W., ANDREW, R. W., Jr., "Radioactivity in bottom sediment of the Clinch and Tennessee Rivers", Disposal of Radioactive Waste into Seas, Oceans and Surface Waters (Proc. Symp. Vienna, 1966), IAEA, Vienna (1966) 57-88.

[22] UKAEA, Analytical Methods for the Determination of Ruthenium-103, Ruthenium-106, Caesium-137, Zirconium-95/niobium-95 and Cerium-144 in Shore Sand and Sea Silt Samples, UKAEA PG Rep. No. 312 (w) (1962).

[23] SODD, V. J., GOLDIN, A. S., VELTEN, R. J., Determination of radioactivity in saline waters, Anal. Chem. 32 (1960) 25-26.

[24] CHAKRAVARTI, D., LEWIS, G. B., PAULUMBO, R. F., SEYMOUR, A. H., Analysis of radionuclides of biological interest in Pacific waters, Nature (London) 203 (1964) 571-73.

[25] YAMAGATA, N., IWASHIMA, K., "Cobalt sulphides, an effective collector for radioruthenium complexes in sea-water", Rapid Methods for Measuring Radioactivity in the Environment (Proc. Symp. Neuherberg, 1971), IAEA, Vienna (1971) 85-90.

[26] SWIFT, E.H., BUTLER, E.A., Precipitation of sulfides from homogeneous solutions by thioacetamide, Anal. Chem. 27 (1956) 146-53.

[27] TSURUGA, H., Occurrence of radioruthenium in the laver, Porphyra tenera, and other seaweeds, Bull. Jap. Soc. Sci. Fish. 28 3 (1962) 372-78.

[28] TSURUGA, H., Sequential analysis of radionuclides in marine organisms, Bull. Jap. Soc. Sci. Fish 31 (1965) 651-58.

[29] TSURUGA, H., Some long-lived radionuclides in marine organisms on the Pacific coast of Japan, J. Radiat. Res. 9 (1968) 63-72.

[30] MERITT, W.F., Radiochemical analysis for long-lived fission products in environmental materials, Can. J. Chem. 36 (1958) 425-28.

[31] KAMBARA, T., Separation of fission products by distillation method. I. Distillation of carrier-free ruthenium tracer with potassium bichromate, Japan Analyst 5 (1956) 222-24.

[32] KAMBARA, T., Studies on the separation of fission products by distillation method. II. Isolation of carrier-free ruthenium by ceric sulfate oxidation, Japan Analyst 6 (1957) 278-80.

[33] LARSEN, R.P., ROSS, L.E., Spectrophotometric determination of ruthenium, Anal. Chem. 31 2 (1959) 176-78.

[34] NAGAYA, Y., SAIKI, M., Accumulation of radionuclides in coastal sediment of Japan (1). Fall-out radionuclides in some coastal sediments in 1964-1965, J. Radiat. Res. 8 1 (1967) 37-43.

[35] TEMPLETON, W.L., PRESTON, A., "Transport and distribution of radioactive effluents in coastal and estuarine waters of the United Kingdom", Disposal of Radioactive Wastes into Seas, Oceans and Surface Waters (Proc. Symp. Vienna, 1966), IAEA, Vienna (1966) 267-89.

[36] KAHN, B., Leaching of some fission products from soil, Anal. Chem. 28 2 (1956) 216.

[37] SHIOZAKI, M., ODA, K., KIMURA, T., SETO, Y., "The artificial radioactivity in sea water", Researches in Hydrography and Oceanography, Commem. Publ. Centenary Hydrographic Department of Japan (1972) 203-49.

[38] HARLEY, J.H., Radiochemical Determination of Ruthenium, USAEC Health and Safety Laboratory Procedures Manual (1972).

[39] KAUTSKY, H., "Possible accumulation of discrete radioactive elements in river mouths", Disposal of Radioactive Wastes into Seas, Oceans and Surface Waters (Proc. Symp. Vienna, 1966), IAEA, Vienna (1966) 163-75.

[40] JONES, R.F., Limnol. Oceanogr. 5 3 (1960) 312.

[41] BELOT, Y., ARCHAMBEAND, M., Comportement du nitrosyl ruthénium sans entraîneur eau de mer simulée, Radioprotection 2 (1967) 163-72.

[42] BELOT, Y., PIGNON, N., Comportement au nitrosyl-ruthénium dans l'eau de surface simulée de chelatants naturels, Health Phys. (1968).

[43] TSURUGA, H., The uptake of radioruthenium by several kinds of littoral seaweed, Bull. Jap. Soc. Sci. Fish. 28 12 (1962) 1149-54.

[44] BHAGAT, S.K., GLYNA, E.P., Nitrosylruthenium-nitro complexes in an aqueous environment, J. Water Poll. Control Fed. 39 (1967) 334-45.

[45] BOVARD, P., "Radioecological field studies carried out in France", Proc. Seminar Marine Radioecology, ENEA, Paris (1969).

[46] WILSON, P.D., Some Aspects of the Chemistry of Ruthenium in Sea Water, PG Rep. 819 (w) (1968).

[47] KOYAMA, M., Nihon Kagaku Zasshi 82 9 (1961) 1182.

[48] WATARI, K., IMAI, K., IZAWA, M., "Coprecipitation behaviour of nitrosylruthenium complexes", Abstracts 14th Ann. Meeting Japan Radiation Res. Soc., J. Radiat. Res. 13 1 (1972) 7.

[49] STROHAL, P., JELISAVCIC, O., The loss of cerium, cobalt, manganese, protactinium, ruthenium and zinc during dry ashing of biological material, Analyst 94 (1969) 678-80.

[50] RILEY, C.J., The Fate of Ruthenium-106 on Ignition of Seaweed (Porphyra) during Analysis, PG Rep. 122 (1962).

[51] NAKAMURA, R., SUZUKI, Y., KAWACHI, E., UEDA, T., The loss of radionuclides in marine organisms during thermal decomposition, J. Radiat. Res. 13 3 (1972) 149-55.

[52] MEADOWS, J.W.T., MATLACK, G.M., Radiochemical determination of ruthenium by solvent extraction and preparation of carrier-free ruthenium activity, Anal. Chem. 34 (1962) 89-91.

[53] WATERBURY, G.H., METZ, C.F., The spectrophotometric determination of alloying and fission product element in non-irradiated plutonium "fissium" alloys, Talanta 6 (1960) 237-45.

DETERMINATION OF RADIONUCLIDES OF Ce, Co, Fe, Ru, Zn AND Zr IN SEAWATER BY PRECONCENTRATION OF COLLOIDAL MANGANESE DIOXIDE

Application to the determination of low-level ruthenium-106

P. GUEGUENIAT, R. GANDON, Y. LUCAS
Commissariat à l'Energie Atomique,
Département de Protection,
Laboratoire de Radioécologie Marine,
Centre de La Hague,
Cherbourg, France

Abstract

DETERMINATION OF RADIONUCLIDES OF Ce, Co, Fe, Ru, Zn AND Zr IN SEAWATER BY PRECONCENTRATION OF COLLOIDAL MANGANESE DIOXIDE. APPLICATION TO THE DETERMINATION OF LOW-LEVEL RUTHENIUM-106.
    General applications of the preconcentration procedure from seawater by means of colloidal manganese dioxide to the determinations of radionuclides of cerium, cobalt, iron, ruthenium, zinc and zirconium are discussed. The conditions for the fixation of ruthenium-106 onto the manganese dioxide was studied in detail, taking the occurrence of various chemical forms of ruthenium-106 in seawater into consideration. The application of this preconcentration procedure to monitoring situations was tested and proved to be satisfactory.

INTRODUCTION

The concentration of certain elements occurring in trace quantities in seawater can be determined by various methods, one of which is coprecipitation. The well-known adsorptive properties of colloidal manganese dioxide [1-3] have been applied to the determination of several metals (stable and radioactive) in seawater.

GENERAL APPLICATION

By using manganese dioxide directly precipitated by the reaction of hydrogen peroxide (30%) with 100 mg potassium permanganate in 2 litres of seawater, the efficiency of fixation of several radionuclides added in a simple chemical form to this medium was measured after 6 hours stirring. The results obtained are illustrated in Fig. 1 as a function of pH. The efficiency of fixation was optimal at the following pH values for various elements treated:

at pH 3.5 for antimony and selenium (90%)
at pH 8 for cerium, cobalt, iron, tin, zinc and zirconium (90-100%)
at pH 10 for cadmium (95%).

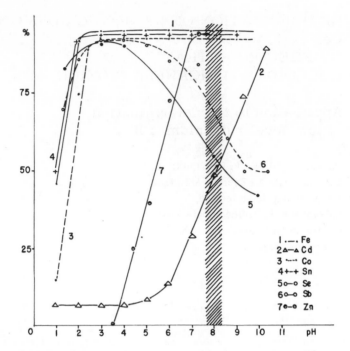

FIG.1.   Variation of efficiency of adsorption of manganese dioxide as a function of pH (pH of seawater is represented as shaded zone).

In addition, it was proved that a fixation of 90-100% can be obtained by treatment at pH 8 for copper, lead and nickel, added to the medium as stable carriers in the concentration range of 100-200 μg/l. This list is not limited to the elements mentioned. It seems that molybdenum [4] in seawater, and yttrium and thorium [5] in freshwater are also adsorbed by manganese dioxide with good yields.

Thus, it is possible to perform determination of the above-mentioned elements in seawater by using this method of preconcentration. The volume of sample water to be treated (using 50 mg of permangante per litre)[1] depends on the concentration of the element to be measured and the sensitivity of the measuring instruments.

In the case of stable element analysis the manganese dioxide is separated by decantation, washed with redistilled water and then dissolved by attack with hot 12$\underline{N}$ HNO$_3$;   elements to be measured are then analysed by atomic absorption spectrophotometry. In treating 2 litres of seawater the precipitate is attacked by 1 ml of 12$\underline{N}$ HNO$_3$ and diluted to 10 ml with redistilled water; the enrichment factor obtained in this condition (200) is sufficient for measuring copper, iron, nickel and zinc occurring in seawater [6]. In the case of radionuclide measurements the manganese dioxide

[1] This quantity can be exceeded advantageously in cases where the manganese dioxide does not interfere with the measurements (e.g. radionuclide determinations).

is separated by filtration after the treatment of a sufficient volume of sea-water, dried in a drying oven at 80°C for 24 hours and then radiological measurements are performed by the aid of γ-spectrometry. Because of the small amounts of the reagents used the background for counting due to manganese dioxide proved to be negligible.

FIXATION OF RUTHENIUM BY MANGANESE DIOXIDE

In the cases mentioned above the elements were added to seawater in a simple chemical form (generally chloride form). Ruthenium, which constitutes one of the essential waste components from irradiated fuel reprocessing plants, occurs in the effluents as complexes, the nature of which can vary from one plant to another, depending on the various treatment procedures used in different plants.

In connection with the problems posed by radioactive waste disposal into seawater, the detection of radioruthenium by preconcentration onto manganese dioxide was examined, taking its physico-chemical behaviour in seawater into account.

Influence of physico-chemical form

As a result of the treatment of irradiated fuels with nitric acid, ruthenium entering as a composition of wastes is found generally in the following forms:

Either in the form of nitrato-complexes of nitrosyl ruthenium $[RuNO(NO_3)_x (OH)_y (H_2O)_z]$ where $x + y + z = 5$, which is unstable and easily hydrolysable in a seawater medium, giving insoluble and soluble forms, the former of which is much more 'adsorbable' than the latter; or in the form of nitro-complexes of nitrosyl ruthenium $[RuNO(NO_2)_x (NO_3)_y (OH)_{z-x-y} (H_2O)_2]$, where $x = 1$, which is stable and little hydrolysable.

It is estimated that the effluents disposed from reprocessing plants into the sea contain essentially nitro-complexes, since the major portion of the nitrato-complexes can be removed by appropriate treatment of the wastes at the plant.

An attempt was made to establish the yield of fixation of various chemical forms of ruthenium in seawater on manganese dioxide[2]. The chemical forms tested were:

(1) Simple forms: ruthenium trichloride
(2) Nitrate-complexes:
    (a) In the initial phase of contact with seawater (mixture of soluble and insoluble forms)
    (b) After 3 weeks' contact with seawater (ruthenium is mainly in soluble complex forms, which can be separated according to the method described previously [7])
(3) Ruthenium in industrial effluents (predominant forms: nitro-complexes).

----

[2] Using 400 mg of KMnO$_4$ per litre of seawater.

TABLE I.   INFLUENCE OF THE PHYSICO-CHEMICAL FORMS OF
RUTHENIUM ON THE EFFICIENCY OF FIXATION OF
MANGANESE DIOXIDE

Forms	RuCl$_3$	Complex nitrato	Soluble forms	Effluent A	Effluent B
Ru fixed (%)	98	95	97	54	98

TABLE II.   FIXATION OF ^{106}Ru ON MANGANESE DIOXIDE UNDER
PROLONGED CONTACT AT ROOM TEMPERATURE

Contact time (d)	Ru fixed (%)	Contact time (d)	Ru fixed (%)
1	55	9	87
2	63	10	90.5
3	67	11	92
4	67	16	94
7	84	18	93
8	87		

The results obtained are shown in Table I. It can be said from these
results that the yields were good (90-100%) for ruthenium trichloride,
nitrato-complexes and the aliquots of the effluents of La Hague reprocessing
plants during the period 1966-1968. On the other hand, for the La Hague
effluent during the period 1972-1973, during which changes in the treatment
of the irradiated fuel were introduced, and for the effluent of the Marcoule
treatment plant the yields were between 40 and 65%.

Improvement of the yield of fixation of nitro-complexes

The relatively low yields obtained by means of manganese dioxide for
ruthenium in the effluent of La Hague in the period 1972-1973 led us to study
more closely the physico-chemistry of ruthenium in water and to take into
account the equilibrium reactions between the various forms present in
order to improve the yield of fixation of ruthenium. In fact, as the result of
disappearance of one or more forms participating in the equilibria in the
solution, the system reacts in such a manner as to reform the adsorbable
forms. Consequently, if the precipitate is left in contact with the solution
to be analysed for a sufficient time, complete fixation of ruthenium on
manganese dioxide may be expected. In addition, certain factors accelerat-
ing this replacement of the equilibria, especially temperature, may be used
advantageously to perform the analysis more rapidly.

TABLE III.   PERCENTAGE OF ADSORPTION OF RUTHENIUM AS A
FUNCTION OF THE AGE OF THE PRECIPITATE

Age of precipitate (d)	Amount adsorbed (%)
0, activity introduced into water before precipitation	58
0, activity introduced into water after precipitation	54
1	53
3	53
5	55
10	52
20	56
35	55

Prolongation of contact time at room temperature

After the fixation of 55% of ruthenium initially present in the medium, the precipitate of manganese dioxide was left to stand in contact with this medium.   The re-establishment of the equilibria was shown by slow increase of adsorption with time.   The results obtained are given in Table II.

Heating

By performing the fixation procedure at 90°C for 3 hours the yield of fixation was about 90% after 1 day of contact time and 90-95% after 3 days.

Comparison of the manganese dioxide method with the cobalt sulphide method

For the fixation of nitro-complexes in seawater Yamagata and Iwashima [8] recommended the use of cobalt sulphide (pH 9.5 at 95°C for 2 hours). By applying this method to the effluent of La Hague (1972-1973) excellent yields, 95-100%, were obtained.

Application of the manganese dioxide method to monitoring

The manganese dioxide method was applied to obtain monthly averages of radioruthenium in seawater in situ.

Technique used

The technique used aimed to obtain one result of total counting at a given time on a cumulative sample which is prepared by stepwise sample collections at a regular time interval and treating each sample immediately after sampling with the same manganese dioxide precipitate. In the present work the duration of each operation was one month and a total volume of seawater of 1000 litres was treated.

The precipitation of manganese dioxide (starting from 0.4 g potassium permanganate per litre) was to begin with performed on 80 litres seawater (the first sampling). After stirring for 6 hours the seawater was left to stand for 12 hours in order to let the precipitate settle and then as much of the supernatant as possible was removed by decantation. The precipitate was preserved at the bottom of the bin used.

A new seawater sample (40-80 litres) was introduced into the bin containing the precipitate, which was used once. The procedures described above, stirring, decantation, etc., were repeated. This operation was repeated as often as necessary, i.e. approximately daily, until 1000 litres of seawater had been treated with the same manganese dioxide precipitate. At the end of the operation the results of the counting on the manganese dioxide allowed one to determine the monthly average concentration of ruthenium at the time considered.

Discussion of the technique

When applying this method to actual operations it is essential that the adsorption properties of the manganese dioxide precipitate do not change in the course of the operation. The adsorptive properties of the precipitate should not change with the age of the precipitate (1 month duration of the operation). If possible, the precipitate used several times in the treatment process should retain the same adsorptive capability as in the first treatment. These points were examined by following series of experiments.

The influence of the age of manganese dioxide precipitate on its adsorptive properties was studied by a series of tests on seawater but using a smaller volume than that used for the monthly measurements, although the same experimental conditions were maintained as when treating 1000 litres of seawater.

First test: Two litres of seawater containing a known activity of ruthenium were treated by precipitating manganese dioxide for adsorption of ruthenium. The precipitate was separated by decantation and filtration and counted for γ-activity.

Second test: Two litres of non-contaminated seawater were used as a medium. Manganese dioxide was precipitated as usual and the same activity of ruthenium as that used in the First test was added after the precipitation procedure. Stirring, filtration and then counting were performed as in the First test.

Third test: The same procedures as for the Second test were used, except that the ruthenium activity was added 24 hours after the precipitation of manganese dioxide.

Other tests: The same cycle of procedures, but the ruthenium activity was added 3, 5, 10, 20 and 35 days after the precipitation of manganese dioxide.

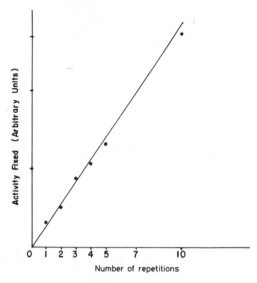

FIG.2.   Variation of efficiency of adsorption of manganese dioxide with repeated precipitations.

The results of these tests are shown in Table III (on nitro-complexes of the effluent of Marcoule plant).   Examination of these results shows that the age of the manganese dioxide precipitate does not play a significant role in the process of adsorption of ruthenium activity.

   Whether there is any change in the adsorption capability of manganese dioxide precipitate through repeated use of the same precipitate was tested by the following experiments.

First experiment: Manganese dioxide was precipitated in 3 litres of seawater containing a known activity of ruthenium.

Second experiment: The same procedures as in the First experiment, but
   after the decantation of the supernatant 3 litres of new seawater containing the same known activity of ruthenium was added to the manganese
   oxide precipitate left on the bottom of the container.   This was then
   processed as usual.

Other experiments: By using identical procedures, the precipitates that
   had been subjected to 3, 4, 5, 7 and 10 adsorption cycles were obtained.
   The results of counting of each precipitate are given in Fig. 2.   This
   figure shows that the adsorbed activity was proportional to the number
   of time the same precipitate was used for the adsorption processes.

   These two series of experiments demonstrate the applicability of the method described to large seawater samples.   The method can also be applied to all elements that are fixed by manganese dioxide.

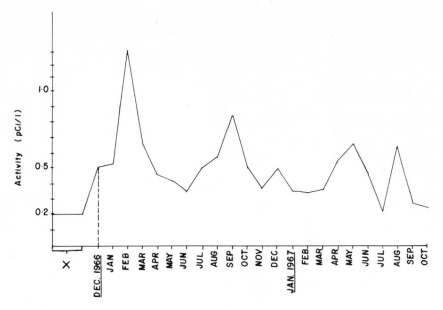

FIG.3.   Monthly activity of ruthenium-106 in seawater from December 1966 to October 1967.   Area X is period before waste disposal operations began in December 1966.

Results obtained at La Hague

The method of ruthenium fixation by manganese dioxide at room temperature was used for a trial within the framework of the monitoring of the sea
area influenced by the waste disposal from the La Hague plant during the
period of 1966-1968 [3, 9].   In this period the yield of fixation of ruthenium
on manganese dioxide was good (90-100%).   The work was carried out
essentially in a limited sector, that of l'Anse d'Ecalgrain, situated 3 km
from the injection point of the pipe-line.   Figure 3 summarizes the average
concentrations (pCi/1) of ruthenium-106 in this sector during this period,
the results of which have already been published elsewhere [7].

To apply this method at present (1972-1973), it is necessary to re-
establish the efficiency of fixation of ruthenium by manganese dioxide and
to take this into account in analysing the results obtained, since the efficiency of fixation has been reduced (40-70%) due to operational changes
in the plant.   To solve this problem two identical 1000-litre seawater samples
were collected, manganese dioxide was precipitated in each sample and the
results obtained by the following treatment were compared to each other:
sample A, measured after 1 day's contact with manganese dioxide (only a
portion of ruthenium was fixed), and sample B, measured after 3 weeks'
contact with manganese dioxide (total ruthenium was fixed).

CONCLUSIONS

The precipitation of manganese dioxide in the seawater medium offers a preconcentration method that is simple and immediately applicable after sample collections. The method can be used for the determination of a number of such trace elements occurring in seawater as cerium, cobalt, copper, iron, nickel, tin, zinc, zirconium, etc. For ruthenium this method can also be used under the following conditions: for small volumes of seawater heat to 90°C to speed up fixation; for large samples this is impractical and the precipitate should be allowed to be in contact with the solution, at room temperature, for three weeks.

REFERENCES

[1] PUSHKAREV, V.V., TKACHENKO, E.V., EGOROV, Yu.V., LYUBIMOV, A.S., Sorption of some radioactive isotopes from aqueous solutions by active manganese dioxide, Radiokhimiya 4 1 (1962) 49.

[2] EGOROV, Yu.V., KRYLOV, E.I., Some characteristics of the sorption of strontium-90 by active manganese dioxide, Radiokhimiya 5 2 (1963) 211.

[3] GUEGUENIAT, P., Détermination de la radioactivité de l'eau de mer en ruthénium, cérium, zirconium par entraînement et adsorption au moyen du bioxyde de manganèse, Rep. CEA-R 3284 (1967).

[4] SUGAWARA, K., TANAKA, M., OKABE, S., Bull. Chem. Soc. Japan 32 (1959) 221.

[5] NOVIKOV, A.I., GORDEEVA, L.N., Coprecipitation of thorium with hydroxides of iron (III), zirconium (IV) and manganese (IV), Radiokhimiya 13 6 (1971) 783.

[6] GUEGUENIAT, P., GANDON, R., "Premiers résultats de mesure de polluants métalliques en baie de Seine et dans le secteur du Cotentin", CIESM XXIIIe assemblée, Athènes, 1972.

[7] GUEGUENIAT, P., Nouvelles études sur les formes insolubles et sur les phénomènes de polymérisation des nitratocomplexes du nitrosylruthénium en eau de mer, C.R. Hebd. Séances Acad. Sci., Paris 274 (1972) 322.

[8] YAMAGATA, N., IWASHIMA, K., "Cobalt sulphides, an effective collector for radioruthenium complexes in sea-water", Rapid Methods for Measuring Radioactivity in the Environment (Proc. Symp. Neuherberg, 1971), IAEA, Vienna (1971) 85-90.

[9] GUEGUENIAT, P., LUCAS, Y., Observations sur la contamination in situ de quelques espèces marines, Note CEA-N-1185 (1969).

RADIORUTHENIUM IN COASTAL WATERS

K. C. PILLAI, N. N. DEY
Health Physics Division,
Bhabha Atomic Research Centre,
Trombay, Bombay,
India

Abstract

RADIORUTHENIUM IN COASTAL WATERS.

The results of observations on ruthenium made on well water and bay water located near the Trombay Fuel Reprocessing Plant are presented. An investigation into the nature of the ruthenium in the well water showed that some 90% of the activity of the water was adsorbed on anion exchange columns and some 10% on cation columns. It was confirmed that only the mobile form of ruthenium reached the well. The distribution of ruthenium in seawater, suspended silt, bottom sediments and in organisms collected from areas close to the effluent discharge is discussed. The forms of ruthenium in seawater were shown to be in two major fractions, 75% in the anionic form and 25% in cationic form. Methods for the determination of radioruthenium are mentioned briefly.

Environmental monitoring in and around nuclear installations releasing radioactive wastes to the environment gives valuable information on the interaction, transport and distribution of radionuclides in the environment consequent on such releases. Some observations made during radiation surveillance at Trombay on radioruthenium in the aquatic environment of the Trombay Bay are discussed here.

The Fuel Reprocessing Plant at Trombay releases the low-level radioactive effluents after treatment to the adjoining Bombay Harbour bay. Major quantities of fission products containing long-lived radionuclides like ^{90}Sr, ^{137}Cs, ^{106}Ru etc. are 'contained' at site. Some freshwater wells are located near the Plant. The wells are monitored regularly for their radioactivity content. During surveillance it was observed that the total beta activity levels in these well waters increased slowly [1]. The frequency of sampling was therefore increased and investigations were carried out on the levels and nature of radioactivity in well waters.

An aliquot of the well-water sample was evaporated to a small volume and subjected to gamma-spectrometric analysis. The gamma spectra indicated the presence of only ^{106}Ru in the sample. The presence of ^{106}Ru was confirmed by beta adsorption and decay measurements. Radiochemical analysis of the well waters showed the presence of ^{90}Sr and ^{137}Cs at very low concentrations. The radioactivity levels in well waters are given in Table I. During the peak activity levels it was observed that more than 99% of the activity of the well waters was contributed by ^{106}Ru. The relative concentrations of ^{137}Cs, ^{90}Sr and ^{106}Ru are given in Table II.

The nature of ruthenium in well waters was investigated. After filtration samples of well waters were passed through cation (Dowex 50×8 - H$^+$ form) and anion (Dowex 1×8 - Cl$^-$ form) columns and the effluents were evaporated and counted separately. The results obtained are given in Table III. It is observed that nearly 94.98% of the activity is anionic.

TABLE I. RADIOACTIVITY LEVELS IN WELL WATERS

Date of collection	Well No.	Total beta activity (pCi/l)	^{106}Ru	^{90}Sr (pCi/l)	^{137}Cs
1968-06-06	1	2 130 ± 75	1771 ± 69	303 ± 9	12.9
1968-06-06	2	2 698 ± 82	2 628 ± 86	-	20.8 ± 15
1969-03-05	1	24 300 ± 185	22 460 ± 120	-	-
1969-06-06	1	25 020 ± 280	21 500 ± 105	45 ± 9	12.1 ± 6.0
1969-06-06	2	985 ± 84	663 ± 30	21 ± 6.5	20.8 ± 5.0
1969-12-06	1	540 ± 42	-	1.55 ± 0.17	2.2 ± 1
1970-05-20	1	1 111 ± 52	-	14.5 ± 6	0.9 ± 0.8
1970-06-15	2	922.3 ± 96	-	20.0 ± 8.0	3.3 ± 2.4
1971-02-08	1	1 803 ± 66	-	7.4 ± 4	1.6 ± 0.5

TABLE II. RELATIVE CONCENTRATION OF ^{137}Cs, ^{90}Sr AND ^{106}Ru IN WELL WATER

Date of collection	Well No.	Relative concentrations		
		^{137}Cs	^{90}Sr	^{106}Ru
1968-06-06	1	1	23.4	137.3
1968-06-06	2	1	-	126.0
1969-06-06	1	1	3.7	1776
1969-06-06	2	1	1.01	31.9
1970-05-20	1	1	16.8	1306
1970-06-15	2	1	6.0	275
1971-02-08	1	1	4.6	1127

Since the total activity of the sample is mostly contributed by radio-ruthenium, it is obvious that most of the ^{106}Ru is present in the anionic form.

Samples of the well waters were dialysed to obtain estimates of the non-dialysable fraction and its ionic nature. After filtration through 0.22 μm Millipore filter paper dialysis was made with 4.8 nm pore diameter dialyser tubing (Arthur Thomas & Co.). The dialysis was carried out for a week till the conductivity of the water was very close to that of the distilled water used. Before dialysis an aliquot of the sample was passed through ion exchange columns. 92.6% of the activity of the water was absorbed on the anion column and 10.5% on the cation column. Only 4% of the total activity was not retained by both cation and anion columns and this was identified as [106]Ru. About 13.5% of the total beta activity remained in the dialyser tubing, which on ion exchange showed that 93.8% was anionic. The non-dialysable fraction in well waters indicates the existence of Ru in colloidal form and/or as organic complexes.

The well water (10 litres) was passed through a soil column (10 g) at a very slow flow rate to study the sorption on the soil column. The soil columns were subsequently separated into three fractions and the beta activity measured. The soil retained only about 4% of the activity, thereby confirming that only the mobile form of Ru had reached the well waters. The ruthenium compounds obtained in the dissolution of irradiated fuels in nitric acid are known to be mainly in the form of nitrosyl derivatives and can exist as cationic, anionic and non-ionized species.

The residue obtained after evaporation of the well water was treated with organic matter extracted from coastal sediments. The filtrate showed that 95% of the solubilized activity was non-cationic and 20% non-anionic, indicating the presence of un-ionized species. About 15% of the solubilized activity was non-dialysable. Earlier Pillai et al. [2] reported that $RuCl_3$, when treated with humic acids, forms non-cationic and/or non-cationic complexes. Species of Ru present in soils (that are not mobile) can eventually get solubilized as organic complexes.

The presence of radioruthenium in well waters near the Fuel Reprocessing Plant site indicates the possibility of the transport of Ru to nearby bay waters.

RADIORUTHENIUM IN BAY WATERS

The distribution of radioruthenium in seawater, suspended silt, bottom sediments and organisms collected from areas close to the effluent-discharge location were studied. Table IV gives the concentration of [106]Ru in seawater and suspended silt. Only very low levels of [106]Ru could be found even in seawaters close to the discharge location. Silt gives high K_d factors for ruthenium and the silt load in the bay waters is high, thereby seawater gets depleted of its ruthenium, which gets deposited locally. Shore and bottom sediments with a high concentration of radioruthenium were confined to the discharge location. Table V gives the concentration of radionuclides in bottom sediments from the discharge location.

Organisms collected from areas close to the discharge location were analysed by gamma spectrometry. Both benthic and non-benthic organisms were collected periodically and analysed. Table VI gives the accumulation of radionuclides in different parts of the various organisms. Only the benthic organisms accumulated both [106]Ru and [144]Ce. Fish flesh accumulated only [137]Cs. It is known that sedimentary matter serves as a source of nutrients for the benthic organisms. The accumulation of radionuclides in sediments also serves as a source of external exposure to such organisms. To compare these two sources of exposure for benthic organisms, computations on the dose delivered from these sources were made for a typical case. Radiation doses received by the organisms from

TABLE III. ION EXCHANGE BEHAVIOUR OF RADIONUCLIDES IN WELL WATERS

Date of collection	Total beta activity (pCi/1)	^{90}Sr (pCi/1)	^{137}Cs (pCi/1)	Cation effluent (pCi/1)	Anion effluent (pCi/1)	Anionic species (%)
1969-04-18	20760 ± 262	-	-	18330 ± 283	432 ± 295	98
1971-02-08	1332 ± 58	7.4 ± 4.0	1.58 ± 0.5	1214 ± 58	53 ± 8	95.5
1970-05-20	1111 ± 52	14.5	0.8	866.2 ± 46	65.3 ± 33	94.0

TABLE IV. DISTRIBUTION OF ^{106}Ru IN SEAWATER AND SILT

Date of collection	Location	Seawater (pCi/l)	Silt (pCi/g (dry))	K_d
1970-04-20	Station 10	1.98 ± 2.8	4.3	2172
1970-06-02	Cirus	1.7 ± 3.0	-	-
1970-06-04	Cirus	5.4	12.36	2296
1970-06-05	TNJ	ND	-	-
1970-06-18	Cirus	0.87 ± 0.15	-	-
1970-12-31	Cirus	0.9 ± 0.6	-	-
1971-03-02	Cirus	ND	-	-
1971-03-18	Cirus	1.44	8.61	5978
1971-03-31	Apollo Pier	ND	-	-
1971-04-12	TNJ	53.1	462.9	8720
1971-11-25	Opp. PP	29.95	166.4	5556
	Opp. PP	398.8	700	1755

potassium-40, ruthenium-106, caesium-137 and cerium-144 in tissues and in sediments are given in Table VII.

Of these nuclides ^{106}Ru contributes most to the exposure of the organism through its accumulation in tissues. However, the contribution from external exposure assumes more significance in contaminated areas.

NATURE OF RADIORUTHENIUM IN COASTAL WATERS

The exact nature of radioruthenium in seawater is not known. Preliminary experiments with filtered seawater showed that about 25% of the radioruthenium exists in cationic form and about 75% in the anionic form. Dialysis of filtered seawater (5 litres) indicated the presence of a small fraction of non-dialysable radioruthenium in seawater. Since the levels of radioruthenium in coastal waters were low, some studies using ruthenium isotopes were carried out to understand the behaviour of Ru in seawater.

When ^{106}RuCl$_3$ was mixed with seawater and allowed to remain in contact it was observed that 60% of the tracer was initially insoluble in seawater (retained on 0.22 μm Millipore filter paper), but progressively more and more Ru was solubilized and at the end of about 44 days only 40% of the original activity was in the insoluble form. The sorption on the Pyrex flask was high about 8% of the total activity. Ion exchange of the filtrate showed partial retention on cation and anion columns, the latter retaining about 85% of the total activity.

An aliquot of the effluents from the Fuel Reprocessing Plant was allowed to interact with filtered seawater (1 : 100 and 1 : 1000) and on

TABLE V. ACCUMULATION OF [106]Ru, ^{137}Cs AND [144]Ce IN BOTTOM
SEDIMENTS AT DISCHARGE LOCATION

Location	[106]Ru	[137]Cs	[144]Ce
	(pCi/g (dry))		
1	116.8	129.3	278.6
2	59.1	89.3	221.9
3	177.5	296.5	538.8
4	59.1	116.1	255.4
5	-	1551.0	313.1
6	3.7	135.7	-
7	-	125.4	97.5
8	-	189.3	391.3
9	37.6	105.2	193.5
10	62.4	149.0	339.5
11	89.0	129.3	234.6

filtering (0.22 μm) it was found that all the ruthenium was in the soluble
form. Ru is found in the anionic state and no non-ionic species were
indicated. However, this might not reflect the actual nature of ruthenium
in seawater, which might possibly change with time since it is known that
nitrosyl ruthenium complexes can undergo substitution reactions in
seawater [3, 4]. Furthermore, removal of ruthenium by silt might alter
the concentration of the various species in seawater.

METHODS FOR DETERMINING RADIORUTHENIUM

The effectiveness of various procedures for the recovery of ruthenium
from environmental samples depends upon the physico-chemical nature of
the sample. Gamma spectrometry without destruction of the sample is
possible only in the absence of interfering nuclides.

Many procedures have been reported for estimation of radioruthenium
[5]. The distillation method is suitable for sediments and organisms.
However, only small samples can be processed by this method. Seawater
samples require a preliminary concentration step to isolate ruthenium
from the bulk of the sample. Alkaline oxidation and subsequent Ru precipitation have been used for water [5]. The procedure carried out with
4 litres of seawater yielded 30-35% recovery only for ruthenium. Ferric
hydroxide [6] and MnO_2 [7] have been used to isolate ruthenium from
large volumes of seawater. Ferric hydroxide precipitation in well waters
yielded only 15-22% recovery of radioruthenium and in seawater 25-65%.

Ruthenium sulphide precipitation using thioacetamide gave 100%
recovery of ruthenium small volumes (100 ml) of well waters and in

TABLE VI. ACCUMULATION OF RADIONUCLIDES IN BENTHIC
AND NON-BENTHIC ORGANISMS FROM THE DISCHARGE LOCATION

Sample	Organism	Parts analysed	^{106}Ru	^{137}Cs	^{144}Ce
				(pCi/g (wet))	
	Benthic organisms				
1	Crab	Flesh	1.44	9.20	1.23
	(Seylla serrata)	Shells	ND	25.68	ND
2		Flesh	5.35	3.71	4.54
		Bones	ND	12.52	ND
		Gut	4.8	1.4	4.6
3		Flesh	20.2	20.7	3.7
		Bones + shell	12.1	35.0	ND
		Gut	73.4	35.5	19.1
4		Flesh	24.7	15.5	7.6
		Bones + shell	85.3	54.7	ND
		Gut	55.8	352.5	35.0
5	Oysters	Flesh	28.5	12.9	71.9
		Shells	ND	60.4	ND
6		Flesh	13.7	8.1	6.9
		Shells	25.3	ND	ND
	Non-benthic organisms				
7	Catfish	Flesh	ND	10.03	ND
	(Arius sp.)	Gut	ND	5.2	ND
8		Flesh	ND	2.56	ND
9		Flesh	ND	29.6	ND
		Bone	ND	62.6	ND
		Gut	230.5	76.2	33.7
10		Flesh	ND	29.1	ND
		Bone	ND	40.1	ND
		Gut	ND	68.6	ND

TABLE VII. RADIATION DOSES RECEIVED BY MARINE ORGANISMS
FROM VARIOUS RADIONUCLIDES IN TISSUES AND IN SEDIMENTS

Nuclide	Dose to the organism from 1 pCi/g of tissue uniformly distributed (mrad/a)	Dose to the organism from 1 pCi/g of sediment (mrad/a)
^{137}Cs	4.43	48.2
^{106}Ru	20.54	30.9
^{144}Ce	20.25	21.3
^{40}K	8.9	13.1

seawaters. However, for larger volumes of seawater the recoveries were reduced to 50% (1 litre) and 30% (4 litres).

Cobalt sulphide [8] has been reported to be an effective collector for all forms of ruthenium complexes in seawater.

ACKNOWLEDGEMENTS

The authors thank Dr. M.R. Iyer for dose calculations. Thanks are due to Shri A. Kuttappan for collection and processing of organisms and Shri R.V. Pawaskar for gamma-spectrometric work.

REFERENCES

[1] DEY, N.N., PILLAI, K.C., "Radioruthenium in aquatic environment of Trombay", Proc. Nat. Symp. Radiation Physics, Bombay (1970).

[2] PILLAI, T.N.V., DESAI, M.V.M., MATHEW, Elizabeth, GANAPATHY, S., GANGULY, A.K., "Organic materials in the marine environment and associated metallic elements", Proc. Symp. Distribution of Chemical Species in Marine Environment, Tokyo (1970).

[3] WILSON, P.F., Some Aspects of the Chemistry of Ruthenium in Sea Water, P.G. Rep. 819(W).

[4] JONES, R.F., The accumulation of nitrosyl ruthenium by fine particles and marine organisms, Limnol. Oceanogr. 5 (1960) 312.

[5] WYATT, E.L., RICKARD, R.R., The Radiochemistry of Ruthenium, NAS. NS-3029 (1961).

[6] KAZUO, W., Concentration of Radionuclides in Sea Water by Ferric Hydroxide Cation Exchange Resin (1966).

[7] GUEGUENIAT, P., Determination of the Ruthenium, Cerium and Zirconium Radioactivity of Sea Water by Carrying over and Adsorption using Manganese Dioxide, Rep. CEA-R-3284 (1967).

[8] YAMAGATA, N., IWASHIMA, K., "Cobalt sulphides, an effective collector for radioruthenium complexes in sea-water", Rapid Methods for Measuring Radioactivity in the Environment (Proc. Symp. Neuherberg, 1971), IAEA, Vienna (1971) 85.

A REVIEW OF METHODS FOR THE
DETERMINATION OF RADIOACTIVE
AND STABLE ZIRCONIUM IN
THE MARINE ENVIRONMENT

J.W.R. DUTTON
Ministry of Agriculture, Fisheries and Food,
Fisheries Radiobiological Laboratory,
Lowestoft, Suffolk,
United Kingdom

Abstract

A REVIEW OF METHODS FOR THE DETERMINATION OF RADIOACTIVE AND STABLE ZIRCONIUM IN THE MARINE ENVIRONMENT.

The origins of the radionuclide of zirconium — from fall-out and radioactive waste disposal — are summarized, together with its nuclear properties. Methods of analysis for this radionuclide are reviewed, especially those suitable for the control of radioactive waste disposal. In this context direct γ-spectrometric analysis of the dried material is generally suitable, but more complex techniques (including radiochemical separations) may be required for ecological research. A summary of methods for the determination of the stable elements in marine materials is also presented.

1. INTRODUCTION

The only important radioactive nuclide of zirconium encountered in the marine environment is ^{95}Zr. This radionuclide is formed during fission with a high yield (about 6%) and its half-life of 65 days results in its persistence in the environment for some time after the other products with shorter half-lives have decayed. In both fall-out and radioactive waste disposal contexts, therefore, it is likely to be encountered with a group of other fission products produced in high yield and with half-lives between 20 and 400 days, viz: ^{141}Ce (33 d), ^{144}Ce (285 d), ^{103}Ru (40 d) and ^{106}Ru (1.0 a). It decays by β^- emission (β_{max} 0.36 and 0.40 MeV), associated with which are 2 γ-photons at 0.724 and 0.757 MeV, which are not in coincidence. The daughter, ^{95}Nb, is also radioactive, with a half-life of 35 days, and undergoes β^- decay, which is also associated with a γ-photon at 0.766 MeV. Zirconium-95 is therefore always associated with some niobium-95.

Zirconium-95 is present in effluent from the Hanford reactors in the USA [1] and in the UK Magnox reactor effluents [2]; it has been measured in water, biota and sediments in the Pacific as a result of waste disposal to the Columbia River [3-5] but has not been detected in the UK as a result of Magnox reactor operation. It is, however, a major component of the discharges from the UK fuel reprocessing plant at Windscale [6] and its appearance around the UK shores, again in water, biota and sediments, has been well documented as a result both of Windscale discharges and smaller releases from Dounreay (on the north Scottish coast) [7-13]. Zirconium-95 has also been measured in algae in part of the La Hague sector affected by waste discharge from the French reprocessing plant at Cap de la Hague [14].

155

TABLE I. ^{95}Zr LEVELS IN SEAFOODS BASED ON
ICRP RECOMMENDATIONS

ICRP public daily maximum intake		1.3×10^5 pCi	
Materials	Fish	Seaweed	Shellfish
Consumption, per day	1 kg	500 g	100 g
DWL	130 pCi/g	260 pCi/g	1300 pCi/g
1% of DWL	1.3 pCi/g	2.6 pCi/g	13.0 pCi/g

Zirconium-95 has been identified and measured in the marine environment due to local fall-out from nuclear-weapon testing [15, 16]. Reports of
its measurement in seawater as a result of stratospheric fall-out have been
summarized up to 1967 by Volchok et al. [17]; other reports include the
measurement of ^{95}Zr in Atlantic Ocean water [18, 19] and surface concentrations in the North Pacific [20]. The measurement of fall-out ^{95}Zr has
also been reported in algae [21] and in other trophic levels [22], in a study
of the oceanic radioecology of ^{95}Zr-^{95}Nb.

2. ANALYSIS FOR ZIRCONIUM-95 IN THE CONTEXT OF RADIOACTIVE WASTE DISPOSAL

UK experience in the control of radioactive discharges indicates that
the major hazard from the disposal of zirconium-95 into the marine environment is the external, whole-body, irradiation of individuals using the beaches
by the ^{95}Zr-^{95}Nb adsorbed onto sands and silts [13, 23][1]. The data indicate
that on the highly adsorbent sediments 1000 pCi ^{95}Zr-^{95}Nb/g (dry) on the
surface with the associated penetration to lower layers will result in a whole-
body dose of 135 μR/h. In the typical situation described, the derived working
limit (DWL) is about 1.7 mrem/h, so the ability of a detection system to
measure 10 pCi/g will enable dose rates of < 0.1% of the DWL to be detected [25].
When considering required lower limits of detection for ^{95}Zr in seafoods,
the International Commission on Radiological Protection (ICRP) value for
^{95}Zr of 1.3×10^5 pCi/d as a maximum acceptable limit for daily intake [26] is
used to obtain a DWL based on the consumption rate of the food. Table I
shows these values, together with 1% of the DWL to indicate a required lower
limit of detection for control purposes.

---

[1] A discussion of the interaction of radionuclides with sediments is given by Duursma and Gross [24].

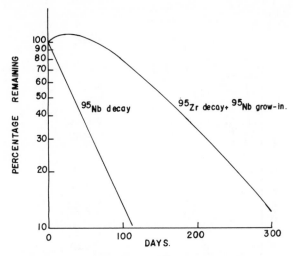

FIG.1.   Decay of ^{95}Zr (with ^{95}Nb grow-in) and ^{95}Nb.

3. THE DETERMINATION OF ZIRCONIUM-95

It has already been noted that ^{95}Zr emits γ-photons at around 0.75 MeV and that the daughter ^{95}Nb has a similar emission. Direct NaI(Tl) γ-spectrometry, which is often used to measure ^{95}Zr-^{95}Nb levels, cannot distinguish between the two radionuclides, and it should be remembered that the method is subject to certain errors due to the variations in ratio of ^{95}Zr:^{95}Nb encountered in different situations. The assumption can be made that the ^{95}Zr and ^{95}Nb are in transient equilibrium (with an activity ratio of 1:2.17) and that the decay between collection and measurement is that of ^{95}Zr alone. If the delay between collection and measurement is long (say 150-200 days), then the ^{95}Zr and ^{95}Nb levels will in any case be approaching those of transient equilibrium; the amount of ^{95}Zr in the sample can then be measured and a decay correction made to the time of sampling based on the ^{95}Zr half-life of 65 days. When total ^{95}Zr-^{95}Nb is required, the sample should be measured as soon as possible after collection to minimize any errors caused by the application of an incorrect decay correction. In spite of the disadvantages noted above, direct NaI(Tl) γ-spectrometric methods have produced much useful information, and will continue to do so.

Figure 1 shows the decay from pure ^{95}Zr and ^{95}Nb sources; the decay curves have been calculated from the equations [27]:

$$^{95}Zr_t = {}^{95}Zr_0\, e^{-\frac{0.693t}{65}}$$

$$^{95}Nb'_t = 2.17\, {}^{95}Zr_0 \left(e^{-\frac{0.693t}{65}} - e^{-\frac{0.693t}{35}} \right)$$

$$^{95}Nb''_t = {}^{95}Nb_0\, e^{-\frac{0.693t}{35}}$$

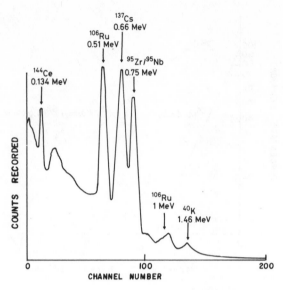

FIG.2.   NaI(Tl) spectrum of Fucus seaweed 0 - 2 MeV.

where $^{95}Zr_0$ and $^{95}Zr_t$ are the ^{95}Zr activities at time 0 and after t days respectively,

$^{95}Nb_0$ is the ^{95}Nb activity at time 0,

$^{95}Nb_t'$ is the ^{95}Nb activity produced by grow-in from ^{95}Zr after t days,

$^{95}Nb_t''$ is the activity after t days left from the original ^{95}Nb activity.

3.1.   The determination of zirconium-95 in seawater

Analytical techniques for this measurement were introduced in the first IAEA reference methods panel in 1970 [28], and reference was made both to laboratory measurements involving separation of ^{95}Zr and ^{95}Nb from seawater followed by γ-counting [29], and to in situ measurements of ^{95}Zr-^{95}Nb [30, 31] using a 100 mm × 80 mm NaI(Tl) crystal and multi-channel analyser. Other published methods of separation or concentration of ^{95}Zr-^{95}Nb from seawater include adsorption onto manganese dioxide, either introduced in the granular form [32] or produced by the oxidation of potassium permanganate with hydrogen peroxide in the sample [33]; these two methods are not, of course, specific for ^{95}Zr-^{95}Nb, and measurement of all adsorbed radionuclides is achieved by γ-spectrometric analysis of the manganese dioxide. Concentration of ^{95}Zr-^{95}Nb from seawater (again with other radionuclides) has been achieved by hydroxide and sulphide precipitations [34], by ion exchange using metal sulphide columns [35] and zirconium hexacyanoferrate [36], and by adsorption onto alumina [37]; evaporation of the seawater has also been used [38, 39].

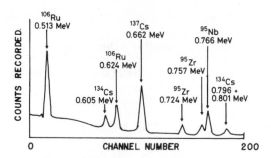

FIG.3.   Ge(Li) spectrum of Fucus seaweed 0.5-0.9 MeV.

All these techniques give good yields for the concentration process, and
most of the reports include details of the subsequent instrumental γ-spectro-
metric analysis of the concentrate, generally using a single NaI(Tl) crystal
detector, although the use of alumina as an adsorbent has been followed by the
multi-dimensional NaI(Tl) γ-spectrometric determination of fall-out ^{95}Zr [40].

A rapid method for the radiochemical determination of ^{95}Zr-^{95}Nb in
seawater has been reported by Ikeda et al. [41], which involves coprecipitation
on La(OH)$_3$ and extraction from 10$\underline{M}$ HCl into 5% tri-n-butyl phosphate in
toluene; high decontamination factors and yields of > 99% were reported. In
recent work at this laboratory [42] the specific separation of ^{95}Zr (and ^{144}Ce)
has been achieved by coprecipitation on Ce, La and Fe hydroxides; after
dissolution with perchloric acid and evaporation to dryness in the presence
of Ru carrier, the residue is dissolved in 11$\underline{M}$ HCl, and Ce and Zr are
separated using an anion exchange column. The Zr is eluted with
9$\underline{M}$ HCl/0.02$\underline{M}$ HF and coprecipitated, with a further addition of La carrier,
as LaF$_3$. The β- or γ-activity of the precipitate is measured and compared
with that of a standard similarly prepared. The yield of the process is
determined by measuring the iron recovered after elution from the column
with 0.5$\underline{M}$ HCl. The method has been used to assess the stability of ^{95}Zr
in seawater from the Windscale discharges, and is currently being used
to measure the distribution of ^{95}Zr in the Irish Sea.

3.2.   The determination of zirconium-95 in biological materials and sediments

An in situ measurement of the contamination of algae by ^{95}Zr-^{95}Nb has
been reported by Guéguéniat [14] but most methods of analysis involve some
processing of the sample and measurement in the laboratory.

The direct γ-spectrometric analysis of the dried or ashed materials
has been reported for the measurement of fall-out ^{95}Zr-^{95}Nb in algae [21]
(after ashing at 500°C) and of ^{95}Zr-^{95}Nb from waste discharge in shore sand
and sea silt [43]. The γ-spectrometric method for the measurement of
^{95}Zr-^{95}Nb in seawater residues [38] is also suitable for the analysis of dried
biological materials and sediments. These methods use 75 mm × 75 mm
NaI(Tl) detectors, but direct γ-spectrometry using Ge(Li) detectors is also in
use in several laboratories (e.g. Refs [44, 45]); these high resolution detectors
enable both ^{95}Zr and ^{95}Nb to be determined, and typical spectra of a Fucus

weed, produced by a 75 mm × 75 mm NaI(Tl) crystal and an 80 cm³ Ge(Li) detector, are shown in Figs 2 and 3.

Several radiochemical separation schemes have been reported for use when ^{95}Zr levels are such that direct γ-spectrometry cannot be used; these schemes (e.g. Refs [41, 47])and the method of Hampson [29] separate ^{95}Zr from other interfering radionuclides.

4. THE DETERMINATION OF STABLE ZIRCONIUM

Reported levels of zirconium in seawater and biological materials are very low indeed, whereas in sediments they are considerably higher; this difference is reflected in the disparity of references to the determination of zirconium in the different materials.

The determination of stable zirconium in seawater has been achieved by spectrophotometry after coprecipitation and ion exchange [48], and by fluorimetry with morin after coprecipitation and solvent extraction [49]. Using this method levels of 0.01 - 0.04 μg/l were obtained in seawater from Japanese coastal areas. It might be expected that neutron activation could be applied to the direct analysis of zirconium in seawater — ^{94}Zr$(n,\gamma)^{95}$Zr and ^{96}Zr$(n,\gamma)^{97}$Zr are possible reactions — but several authors have noted interference due to fission of uranium and consequent production of ^{95}Zr [50, 51]. The use of neutron activation, therefore, is limited to irradiation after separation.

The determination of zirconium in ashed samples of marine animals and plants has been reported using fluorimetry [49] and spectrophotometry with Alizarin Red S [52].

Several instrumental techniques have recently been used to determine zirconium in sediments and sands, including X-ray fluorescence (e.g. Refs [53, 54]) and emission spectrography (e.g. Refs [55, 56]); Ref.[55] includes details of a preliminary separation of Zr, Nb, Ta and Hf by precipitation with benzene arsonic acid. Several neutron activation reports have been noted, including the use of 14 MeV neutrons to study trace elements in silicates with a sensitivity of 10-100 μg Zr [57], and a sequential scheme has been published to measure zirconium among 30 other elements in marine deposits [58]. Spectrophotometry has also been used to measure zirconium in sediments using Alizarin Red S [52] and in silicate rocks using quinalizarin sulphonic acid, with twice the sensitivity of Alizarin Red S [59].

5. CONCLUSIONS

Although standard NaI(Tl) γ-spectrometric analysis of marine materials is generally suitable for the determination of ^{95}Zr(-^{95}Nb) in a control context, the use of Ge(Li) detectors will improve the quality of the data by producing separate ^{95}Zr and ^{95}Nb results; this is true also for radioecological investigations, especially in a regime of waste control discharges. For studies of fall-out ^{95}Zr(-^{95}Nb) in the marine environment reliance must still be placed on NaI(Tl) spectrometry, which for multinuclide analysis of seawaters involves either the processing of large volumes of sample, or the use of complex detection systems, or both. Recourse may have to be made to the separation of ^{95}Zr alone.

Very few stable zirconium data have been reported for the seawater/biota system due to the low levels of the element in the environment and the consequent difficulty of analysis. Most investigations into marine biological processes have been carried out by studying the radionuclide ^{95}Zr.

REFERENCES

[1]   PARKER, H.M., in Industrial Radioactive Waste Disposal Hearings before the Joint Committee on Atomic Energy 1, US Government Printing Office, Washington, D.C. (1959) 202-41.
[2]   DUTTON, J.W.R., HARVEY, B.R., Water Res. 1 (1967) 743-57.
[3]   SEYMOUR, A.H., LEWIS G.B., USAEC Rep. UWFL-86 (1964).
[4]   FORSTER, W.O., ROBERTSON, D., Rep. RLO-1750-54 (1969) 67-70.
[5]   OSTERBERG, C., KULM, L.D., BYRNE, J.V., Science 139 (1963) 916-17.
[6]   HOWELLS, H., in Disposal of Radioactive Wastes into Seas, Oceans and Surface Waters (Proc. Symp. Vienna, 1966), IAEA, Vienna (1966) 769-85.
[7]   MAUCHLINE, J., UKAEA Rep. AHSB(RP) R 27 (1963).
[8]   ANON, Radioactivity in Surface and Coastal Waters of the British Isles, Ministry of Agriculture, Fisheries and Food, Lowestoft, Tech. Rep. FRL.1 (1967).
[9]   MITCHELL, N.T., Radioactivity in Surface and Coastal Waters of the British Isles 1967, Ministry of Agriculture, Fisheries and Food, Lowestoft, Tech. Rep. FRL.2 (1968).
[10] MITCHELL, N.T., Ibid. for 1968, FRL.5 (1969).
[11] MITCHELL, N.T., Ibid. for 1969, FRL.7 (1971).
[12] MITCHELL, N.T., Ibid. for 1970, FRL.8 (1971).
[13] JEFFERIES, D.F., Helgoländer Wiss. Meeresunters. 17 (1968) 280-90.
[14] GUEGUENIAT, P., Bull. Inf. Sci. Tech. (Paris) 151 (1970) 23-26.
[15] SEYMOUR, A.H., HELD, E.E., LOWMAN, F.G., DONALDSON, J.R., SOUTH, D.J., USAEC Rep. UWFL-47 (1957).
[16] MIYAKE, Y., SUGIURA, Y., Rec. Oceanog. Works Japan 2 (1955) 108-12.
[17] VOLCHOK, H.L., BOWEN, V.T., FOLSOM, T.R., BROECKER, W.S., SCHUERT, E.A., BIEN, G.S., Radioactivity in the Marine Environment, National Academy of Sciences (1971) Chap. 3.
[18] YOUNG, J.A., Pacific Northwest Laboratory Ann. Rep. for 1968, BNWL-1051, Pt. 2 (1969) 104-8.
[19] SILKER, W.B., RIECK, H.G., ibid., 32-35.
[20] SILKER, W.B., ibid., 27-32.
[21] ANGINO, E.E., SIMEK, J.E., DAVIS, J.A., Pub. Inst. Marine Sci. Univ. Tex. 10 (1965) 173-78.
[22] OSTERBERG, C., PEARCY, W.G., CURL, H., Jr., J. Mar. Res. 22 (1964) 2-12.
[23] JEFFERIES, D.F., Environmental Surveillance in the Vicinity of Nuclear Facilities (REINIG, W.C., Ed.), Charles C. Thomas, Springfield, Illinois (1970) 205-16.
[24] DUURSMA, E.K., GROSS, M.G., Radioactivity in the Marine Environment, National Academy of Sciences (1971) Chap. 6.
[25] JEFFERIES, D.F., personal communication.
[26] INTERNATIONAL COMMISSION ON RADIOLOGICAL PROTECTION, Report of Committee II on Permissible Dose for Internal Radiation (1959), ICRP Publication No. 2, Pergamon Press, Oxford (1960).
[27] OVERMAN, R.T., CLARK, H.M., Radioisotope Techniques, McGraw-Hill, New York (1960) Chap. 8 and App. H.
[28] CHESSELET, R., Reference Methods for Marine Radioactivity Studies, Technical Reports Series No. 118, IAEA, Vienna (1970) 275-84.
[29] HAMPSON, B.L., Analyst 88 (1963) 529-33.
[30] CHESSELET, R., LALOU, C., NORDEMANN, D., Symp. Contamination of the Marine Environment, C.E.R.B.O.M., Nice (1964).
[31] CHESSELET, R., Rep. CEA-R-3698 (1968).
[32] YAMAGATA, N., IWASHIMA, K., Nature (London) 200 (1963) 52.
[33] GUEGUENIAT, P., Rep. CEA-R-3284 (1967).
[34] CHAKRAVARTI, D., LEWIS, G.B., PALUMBO, R.F., SEYMOUR, A.H., Nature (London) 203 (1964) 571-3.
[35] WATARI, K., TSUBOTA, H., KOYANAGI, T., IZAWA, M., Nippon Genshiryoku Gakkai-Shi 8 (1966) 182-5.
[36] KAWAMURA, S., SHIBATA, S., KUROTAKI, K., Anal. Chim. Acta 56 (1971) 405-13.

[37] SILKER, W.B., RIECK, H.G., Pacific Northwest Laboratory Ann.Rep. for 1968, BNWL-1051, Pt.2
      (1969) 18-20.
[38] DUTTON, J.W.R., Ministry of Agriculture, Fisheries and Food, Lowestoft, Tech.Rep.FRL.4 (1969).
[39] CHESSELET, R., LALOU, C., Radioactivity in the Sea No.13, IAEA, Vienna (1965).
[40] PERKINS, R.W., ROBERTSON, D.E., RIECK, H.G., Pacific Northwest Laboratory Ann.Rep. for 1965,
      BNWL-235, Pt.2 (1966) 108-15.
[41] IKEDA, N., KIMURA, K., IZAWA, K., YASUI, T., Radioisotopes (Tokyo) 18 (1969) 8-11.
[42] DUTTON, J.W.R., IBBETT, R., unpublished work.
[43] UNITED KINGDOM ATOMIC ENERGY AUTHORITY, P.G. Rep. 312 (1962).
[44] DUTTON, J.W.R., unpublished work, 1968.
[45] COOPER, J.A., WOGMAN, N.A., PALMER, H.E., PERKINS, R.W., Health Phys. 15 (1968) 419-33.
[46] BONI, A.L., Anal.Chem. 32 (1960) 599-604.
[47] TSUMGA, H., Bull.Japan Soc.Sci.Fish. 31 (1965) 651-58.
[48] HEDGES, D.H., HOOD, D.W., in The Chemistry and Analysis of Trace Metals in Sea Water
      (Hood, D.W., Ed.), TID-23295.
[49] SHIGEMATSU, T., NISHIKAWA, Y., HIRAKI, K., Nippon Kagaku Zasshi 85 (1964) 490-93.
[50] SCHUTZ, D.F., TUREKIAN, K.K., Geochim.Cosmochim. Acta 29 (1965) 259-313.
[51] PERKINS R.W., ROBERTSON, D.E., Rep. BNWL-SA-1007.
[52] SASTRY, V.N., KRISHNAMOORTHY, T.M., SARMA T.P., Curr.Sci. (India) 38 (1969) 279-81.
[53] BOHDAN, D., HOLYNSKA, B., STACHURSKI, J., Chem.Anal. (Warsaw) 12 (1967) 1107-11.
[54] HELLMANN, H.Z., Anal.Chem. 254 (1971) 192-95.
[55] TSYKHANSKII, V.D., KRINBERG, I.A., Izv.Sibirsk.Otd.Akad.Nauk SSSR 11 (1963) 125-27.
[56] MEISHTAS, E.A., Okeanologiya 10 (1970) 373-82.
[57] PERDIJON, J., Bull.Soc.Franc.Ceram. 67 (1965) 45-51.
[58] LANDSTROM, O., SAMSAHL, K., WENNER, C.G., Proc.Int.Conf.Modern Trends in Activation Analysis,
      Gaithersburg, Md., (DeVOE, J.R., Ed.) 1, N.B.S., Washington (1969) 353-66.
[59] CULKIN, F., RILEY, J.P., Anal.Chim.Acta 32 (1965) 197-210.

A REVIEW OF METHODS FOR THE DETERMINATION OF RADIOACTIVE AND STABLE SILVER IN THE MARINE ENVIRONMENT

J.W.R. DUTTON
Ministry of Agriculture, Fisheries and Food,
Fisheries Radiobiological Laboratory,
Lowestoft, Suffolk,
United Kingdom

Abstract

A REVIEW OF METHODS FOR THE DETERMINATION OF RADIOACTIVE AND STABLE SILVER IN THE MARINE ENVIRONMENT.

The origins of the radionuclide of silver — from fall-out and radioactive waste disposal — are summarized, together with its nuclear properties. Methods of analysis for this radionuclide are reviewed, especially those suitable for the control of radioactive waste disposal. In this context direct γ-spectrometric analysis of the dried material is generally suitable, but more complex techniques (including radiochemical separations) may be required for ecological research. A summary of methods for the determination of the stable elements in marine materials is also presented.

1. INTRODUCTION

Nuclides of silver from atomic weight 105 to 114 inclusive are shown in Table I, together with their half-lives, modes of decay and production processes. The list includes all those radionuclides that might conceivably be encountered in the marine environment as a result of nuclear weapon test fall-out, radioactive waste disposal and labelled tracer studies.

Radiosilver was first recorded in the marine environment by Seymour [5], who reported the presence of $^{110}Ag^m$ (from the Hardtack test series at Eniwetok in 1958) in the liver of a spiny lobster collected in 1959 at Guam. In 1964 Folsom and Young [6] investigated the distribution of $^{110}Ag^m$ in oceanic and coastal organisms of the Pacific and found it to be widespread, due to fall-out and water movement. The radionuclide has since been measured in the Pacific Ocean in salmon [7] and tuna [8], in particulate material [9], and in seaweeds [10].

Silver-110m is present in effluents from Magnox-type reactors [11], and in liquors from the British Nuclear Fuels Ltd. irradiated fuel reprocessing plant at Windscale [12]. The presence of this radionuclide is probably due to the neutron activation of ^{109}Ag, but whether this stable isotope is present as an impurity in the Magnox cladding (or the uranium) or whether it is formed as the stable terminator of the short-lived fission chain $^{109}Rh \rightarrow ^{109}Pd \rightarrow ^{109}Ag$ has not been established. The detection of $^{110}Ag^m$ as a result of discharges from the above installations was reported in 1968 [13], and radioecological data have been obtained by studying its movement through the environment in the Irish Sea and the Blackwater River Estuary [14]. The operations of USA plant, e.g. at Hanford, have resulted in the appearance of $^{110}Ag^m$ in the Pacific Ocean [7,8,10] in various materials.

TABLE I. NUCLIDES OF SILVER [1-4]

Nuclide	105	106m/106	107	108m/108	109	110m/110	111m/111	112	113m/113	114
Half-life	40 d	8.2 d/24 min	stable (51.35%)	127 a/2.3 min	stable (48.65%)	253 d/24 s	1.2 min/7.6 d	3.2 h	1.2 min/5.3 h	5 s
Decay process	EC	$\beta^+ \beta^-$ EC	stable {107 m: 44.3 s EC} IT	EC IT/β^- (β^+, EC)	{109 m: 39 s EC} IT	β^-, IT/β^-	IT/β^-	β^-	β^-/β^-	β^-
γ-photon energies (MeV)	0.345 0.281 0.064 others	0.51 1.05 1.53 others	-	^{108}Ag^m 0.434 – 89% 0.614 – 90% 0.722 – 90%	-	^{110}Ag^m 0.656 – 96% 0.884 – 71% 0.937 – 32% others	0.342 0.247	0.618 1.39 others	0.14 0.30 others	0.157
Production process	Complex	Complex	-	^{107}Ag (n, γ) ^{109}Ag (n, 2n) ^{108}Cd (n, p)	-	^{109}Ag (n, γ)	fission, yield: 0.015%	0.012%	0.013%	0.011%

From thermal neutron fission of ^{235}U, fast fission yield will be greater

Because $^{110}Ag^m$ was not a radionuclide caused by fission in any significant quantity, Folsom and Young [6] suggested that it had been produced by neutron activation of stable ^{109}Ag in the nuclear device, and looked further for and found $^{108}Ag^m$ produced presumably from the other stable isotope ^{107}Ag; the $^{110}Ag^m$:$^{108}Ag^m$ ratio was then used to follow the movement of oceanic water masses [15]. There has since been discussion [16, 17] on the possible production processes — (n,γ) and (n, 2n) reactions with stable silver isotopes, and (n,p) reactions with isotopes of stable cadmium — but all workers are hopeful that this method of dating will prove useful. Silver-108m has also been reported in tuna [8] and salmon [18] in addition to the original levels reported in squid, and the spiny lobster [15].

The deliberate introduction of radionuclides as tracers into the marine environment has become more widespread in the past few years, and, because silver is strongly adsorbed onto sludges, $^{110}Ag^m$ is finding an increasing use in this field [19]. The half-life, 253 days, is long enough to allow an extended study, and the γ-photon emissions at 0.66 and 0.89 MeV permit easy detection.

2. ANALYSIS FOR SILVER-110m IN THE CONTEXT OF THE CONTROL OF RADIOACTIVE WASTE DISPOSAL

Committee 2 of the International Commission on Radiological Protection (ICRP) has recommended [20] maximum permissible body burdens and equivalent maximum permissible concentrations in drinking water, which are consistent with the recommendations of the main Commission regarding dose limits. By using these concentrations and rates of consumption in edible marine materials it is possible to derive maximum acceptable levels of $^{110}Ag^m$ in these materials. These are known as derived working limits (DWL) and examples are shown in Table II, together with 1% of these derived limits to indicate a required lower limit of detection for control purposes.

Concentration factors[1] for many elements in marine biota have been summarized recently [21-23]: the data on silver for seafoods are quoted from this report in Table III. These data can be used to determine the concentrations of $^{110}Ag^m$ in seawater that should produce 1% of the DWL in the seafoods; these concentrations (based on the highest concentraction factor reported) are regarded as the required sensitivity of detection and are also shown in Table III.

In fact, the determination of $^{110}Ag^m$ in seawater is not required for control purposes because it is always easier to measure the level of the radionuclide in the edible material. No measurements of $^{110}Ag^m$ – or $^{108}Ag^m$ – in seawater have, in fact, been reported.

3. THE DETERMINATION OF RADIONUCLIDES OF SILVER

Only $^{108}Ag^m$ and $^{110}Ag^m$ are likely to be encountered in the marine environment, the other radionuclides being either unusual or, if produced by fission, of very short half-life. $^{110}Ag^m$ is of importance in the context of

---

[1] The term 'concentration factor' is defined as 'concentration of element per unit weight in material divided by concentration of element in the same weight of seawater'.

TABLE II. $^{110}Ag^m$ LEVELS IN EDIBLE SEAFOODS
BASED ON ICRP RECOMMENDATIONS

ICRP public daily maximum intake	6.6×10^4 pCi		
Material	Fish	Seaweed	Shellfish
Consumption per day	1 kg	500 g	100 g
DWL	66 pCi/g	132 pCi/g	660 pCi/g
1% of DWL	0.7 pCi/g	1.3 pCi/g	6.6 pCi/g

TABLE III.   REQUIRED SENSITIVITY OF DETECTION
OF $^{110}Ag^m$ IN SEAWATER

Material	Fish	Seaweed	Shellfish
1% of maximum level in material	0.7 pCi/g	1.3 pCi/g	6.6 pCi/g
Concentration factor	≈ 10	$10^2 - 10^3$	$10 - 10^4$
Corresponding level in seawater	70 pCi/l	1.3 pCi/l	0.7 pCi/l

control of waste disposal; both are important to environmental research
and investigation.

Reference to the γ-photons emitted by the two nuclides, which are shown
in Table I, shows that both $^{108}Ag^m$ and $^{110}Ag^m$, if present in sufficient
quantities, could be measured by γ-spectrometric analysis. Most of the
reported data on the two radionuclides have been based on some form of
γ-spectrometry, using chemical separations mainly to characterize and
confirm the presence of the nuclide. Thus, the original detection of $^{110}Ag^m$ [5]
in the spiny lobster at Guam was by simple γ-spectrometry of the ashed
sample, using a 75 mm $\times$ 75 mm NaI(Tl) crystal and 256-channel analyser
to view the 0.66 and 0.89 MeV photopeaks; the detection was confirmed by
Palumbo and co-workers by chemical separation. Folsom and Young [6]
determined $^{110}Ag^m$ using a triple-γ summing-coincidence technique. The
radionuclide emits 3 γ-photons in coincidence, at 0.656, 0.884 and 1.384 MeV,
which produce the triple coincidence peak at 2.92 MeV; the count rate in this
peak was measured using two 200 mm $\times$ 100 mm NaI(Tl) crystals spaced
25 mm apart, with the ashed or dried material placed between them. The
use of the summing-coincidence technique enabled $^{110}Ag^m$ to be measured
with negligible interference from other photon emissions.

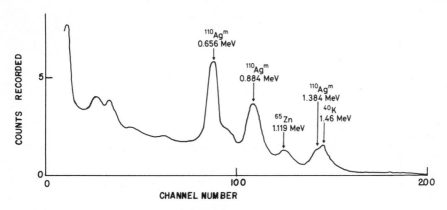

FIG.1.   Gamma spectrum of dried oyster sample (counts recorded are in a relative unit).

Beasley and Held [16] used both NaI(Tl) and Ge(Li) γ-spectrometric systems to measure [110]Ag[m], as did Robertson et al. [9]. The measurement of [110]Ag[m] by United Kingdom workers [13], in the presence of relatively large amounts of other fission-product and neutron-activated radionuclides, was accomplished by NaI(Tl) spectrometry with a standard 75 mm × 75 mm crystal, using both the matrix method [24] and an iterative least-squares fit technique [25] to resolve the spectrum, a typical example of which is shown in Fig.1; the presence of [110]Ag[m] was again confirmed by chemical separations. Silver-110m levels in marine materials are now being regularly reported in marine materials around the UK shores [26-29], and are estimated by the above method.

Silver-108m was measured by Folsom and Young [6] initially using the summing-coincidence technique, and later a dual-parameter NaI(Tl) spectrometer. Beasley and Held [16] obtained additional [108]Ag[m] data by NaI(Tl) and Ge(Li) spectrometry after radiochemical separation of silver from other interfering radionuclides. Cooper et al. [18] used an anticoincidence-shielded dual Ge(Li) detector system.

If the direct measurement of [110]Ag[m] (or [108]Ag[m]) is impracticable due either to their low levels or to relatively high levels of other interfering radionuclides, then recourse must be made to chemical separations. The published sequential schemes (e.g. Refs [30-32]) for radionuclide separation from biological materials do not include a radiosilver separation stage, but there are several schemes for the sequential separation of radionuclides including [110]Ag[m] from neutron-activated [33] and liquid effluent [11] samples; both types of scheme could be modified either by the addition of radiosilver or for the larger sizes of environmental samples.

Details of specific methods for radiosilver separation are given in the Nuclear Science Series monograph The Radiochemistry of Silver [34]; pre-cipitation of AgCl (used by the UK workers [13]) and Ag_2S, coprecipitation with Hg_2Cl_2, solvent extraction as the dithizone complex, ion exchange using cation and anion resins, isotopic exchange using solid AgCl, and electro-deposition — all are mentioned in the review with an extensive bibliography.

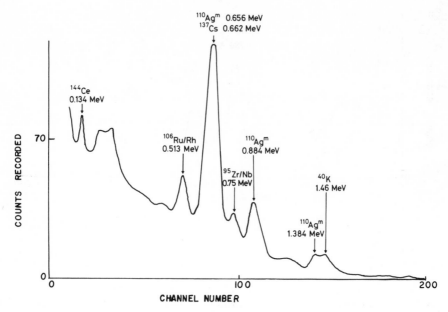

FIG.2.   Gamma spectrum of dried sediment sample (counts recorded are in a relative unit).

More recently Beasley and Held [16] in their chemical separation of radio-silver used a trialkyl thiophosphate as a selective extractant [35]; Hodge and Folsom [36] have published a procedure for the direct removal of radiosilver from kilogram quantities of wet tissue by complexing with cyanide containing 1 gram of silver. The silver can then be plated onto platinum, redissolved in nitric acid and finally precipitated as silver chloride for subsequent γ-spectrometric analysis.

For pure, separated, sources of $^{110}Ag^m$ other counting techniques available besides γ-spectrometry are β-counting of an electrodeposited source or gel scintillation counting of, for example, the white silver chloride precipitate.

The determination of $^{110}Ag^m$ in sediments is generally required as a result of tracer experiments, but may also be required as a result of controlled disposal of liquid radioactive wastes. In both these cases direct γ-spectrometry by NaI(Tl) detectors should be adequate, either in situ [37] or laboratory based. Figure 2 shows the spectrum of a silt, sampled in Liverpool Bay in the Irish Sea as part of a tracer study and analysed at this laboratory. The $^{110}Ag^m$ photopeaks are overlapped by fission-product peaks of ^{137}Cs, ^{134}Cs and $^{95}Zr/^{95}Nb$ present in low concentrations from the Windscale discharges.

When selecting an analytical technique, several considerations must be borne in mind, such as the sensitivity and selectivity of the method and its ease and reliability of operation, whether a multi-element capability is required, and the capital and running costs of the system. In an attempt to compare the performance of the various counting systems used to measure $^{108}Ag^m$ and $^{110}Ag^m$, data on their reported performance have been collated and are shown in Table IV.

TABLE IV.   REPORTED PERFORMANCE OF COUNTING SYSTEMS

System	pCi in sample	
	$^{108}Ag^m$	$^{110}Ag^m$
Direct 75 mm × 75 mm NaI(Tl) [23]	-	50 ± 10[a]
γ-summing spectrometer 200 mm × 100 mm NaI(Tl) detectors [6]	0.2 ± 0.2[b]	10 ± 5[c]
2-dimensional shielded γ-spectrometer 380 mm × 280 mm NaI(Tl) detectors [15]	1.07 ± 0.17[d]	-
Anticoincidence-shielded dual Ge(Li) 70 × 80 cm³ detectors [18]	5 ± 0.5[e]	-

[a] 2σ obtained by least-squares fit on oyster sample.
[b] Counting standard error in the presence of 20 pCi $^{110}Ag^m$.
[c] Lowest $^{110}Ag^m$ reported; squid liver sample, assumed weight 50 g, 1000-minute count.
[d] One sigma standard deviation of counting, including background effect.
[e] Accuracy in the presence of 500 pCi ^{40}K.

Unfortunately, the figures given are not directly comparable because of
the wide difference in other activities present in the sample and the termin-
ology used by the analysts. There is no doubt, however, that direct
γ-spectrometric analysis of the dried material with a 75 mm × 75 mm NaI(Tl)
detector and a 200-channel pulse height analyser can successfully be applied
in the analysis of the marine environment for $^{110}Ag^m$ (as well, of course, as
many other radionuclides) in a radioactive waste disposal context, and that
there is generally no need for radiochemical separations. More sophisticated
detector assemblies, using larger NaI(Tl) crystals operated in some coincidence
counting mode, or large Ge(Li) detectors (volume perhaps 80 cm³) may be
required to measure the levels of $^{110}Ag^m$ encountered in environmental
research, and certainly those due to $^{108}Ag^m$.

4.   THE ANALYSIS OF STABLE SILVER

4.1. The determination of silver in seawater

Methods for the determination of silver in seawater employed up to 1963
have been summarized by Riley [38]: concentration before determination
involving coprecipitation with mercuric sulphide [39], cocrystallization with
thioanilide [40] and tannin [41], anion exchange [42], solvent extraction [43]
and freeze-drying [44]. Determination of the concentrated silver was carried
out by emission spectrography [39, 41], photometry [40, 43] and neutron
activation, using the reaction $^{109}Ag(n,\gamma)^{110}Ag^m$, followed by chemical separation
of the $^{110}Ag^m$ and NaI(Tl) γ-spectrometry [44]. Levels from < 0.1 to 3 μg/l
were reported by these investigators.

More recently analytical techniques reported include neutron activation followed by chemical separation and Ge(Li) γ-spectrometry [45], solvent extraction followed by emission spectrographic analysis of the extract [46] and solvent extract — using ammonium pyrollidine dithiocarbamate (APDC)/ chloroform — followed by atomic absorption spectrophotometry (AAS) using a tantalum boat [47]. In a recent trace-element intercomparison study [48] one of the participants measured silver in seawater by anodic stripping voltammetry (with a carbon electrode). Levels of silver from about 0.01 to 1 μg/l were reported by these investigators.

It seems that most data on silver in seawater have been obtained by neutron activation (e.g. Refs [44, 45, 49]), but the use of solvent extraction/ AAS techniques, already applied to the determination of many transition elements in seawater (e.g. Refs [50, 51]), may soon be extended to produce more data than the single reference already mentioned.

4.2. The determination of silver in biological materials and sediments

The techniques used for the determination of silver in marine biota and sediments are basically the same as for seawater. Emission spectrography has been used to detect or measure silver in plankton [52], seaweeds [41] and sediments [53]. Neutron activation has been applied to the determination of silver in fish by a direct instrumental method [54] and in sediments after chemical separation [55]. Silver has also been determined in seaweeds and molluscs by direct AAS on the dissolved materials [47], and in sediments after extraction of the element with trioctyl thiophosphate in methyl isobutyl ketone [56].

The use of the tantalum boat in AAS analysis for silver in soils has also been reported [57] and this technique with its very low limit of detection (about 0.1 ng) could perhaps be applied to sediments also.

In all these methods it is important to use standard reference materials to establish the validity of the analytical techniques. The determination of silver in a kale sample [58] by several workers has shown some variations in results: a figure of about 0.4 ppm has been obtained by AAS [47] and neutron activation [59, 60], but other workers [61] have obtained < 0.01 ppm by neutron activation. They have suggested that inhomogeneity may be the cause of this discrepancy. Whatever the reason, more intercomparison work between methods and laboratories is very necessary.

5. CONCLUSIONS

Some form of γ-spectrometry, using either NaI(Tl) or Ge(Li) detectors, is the preferred technique for the analysis of $^{110}Ag^m$ and $^{108}Ag^m$ in the context of both waste control and research, with the simpler assemblies (e.g. 75 mm × 75 mm NaI(Tl) crystal and 200-channel pulse-height analyser) being adequate for most control situations. Chemical separations are generally not necessary, but can be used if required to increase sensitivity and decrease interference. The determination of stable silver in the marine environment is effected by several techniques, with neutron activation the most popular. The intercomparability of the techniques, and of the laboratories participating in both waste control and environmental research, is of prime importance.

REFERENCES

[1]   FEDERAL MINISTRY OF SCIENTIFIC RESEARCH, Chart of the Nuclides, Bad Godesberg, 2nd ed. (1963).
[2]   LEDERER, C.M., HOLLANDER, J.M., PERLMAN, I., Table of Isotopes, Wiley, New York, 6th ed. (1967).
[3]   CROUTHAMEL, C.E., Applied Gamma-Ray Spectrometry, Pergamon Press, Oxford, 2nd ed. (1970) pp. 752.
[4]   HARBOTTLE, G., Radiochim. Acta 13 (1970) 132-34.
[5]   SEYMOUR, A.H., in Radioecology (SCHULTZ, V., KLEMENT, A.W., Jr., Eds), Reinhold, New York,
      and Am.Inst.Biol.Sci. (1963) 151-57.
[6]   FOLSOM, T.R., YOUNG, D.R., Nature (London) 206 (1965) 803-6.
[7]   JENKINS, C.E., Health Phys. 17 (1969) 507-12.
[8]   FOLSOM, T.R., YOUNG, D.R., HODGE, V.F., GRISMORE, R., in Third National Symposium on
      Radioecology, CONF-710501-33 (1971).
[9]   ROBERTSON, D.E., RANCITELLI, L.A., PERKINS, R.W., in Proc.Int.Symp.Applications of Neutron
      Activation Analysis in Oceanography, Brussels (1968) 143-212.
[10] WONG, K.M., HODGE, V.F., FOLSOM, T.R., in Plutonium Symposium, CONF-7180817-1 (1971).
[11] DUTTON, J.W.R., HARVEY, B.R., Water Res. 1 (1967) 743-57.
[12] DUTTON, J.W.R., WOODMAN, F.J., personal communication, 1968.
[13] PRESTON, A., DUTTON, J.W.R., HARVEY, B.R., Nature (London) 218 (1968) 689-90.
[14] PRESTON, A., JEFFERIES, D.F., MITCHELL, N.T., Nuclear Techniques in Environmental Pollution
      (Proc.Symp.Salzburg, 1970), IAEA, Vienna (1971) 629-44.
[15] FOLSOM, T.R., GRISMORE, R., YOUNG, D.R., Nature (London,227 (1970) 941-43.
[16] BEASLEY, T.M., HELD, E.E., Nature (London) 230 (1971) 450-51.
[17] GRISMORE, R., FOLSOM, T.R., YOUNG, D.R., Nature (London) 234 (1971) 347.
[18] COOPER, J.A., PERKINS, R.W., KOSOROK, J.R., JENKINS, C.E., Pacific Northwest Laboratory Ann.
      Rep. for 1970, BNWL-1551: Pt.2 (1971) 54-55.
[19] WHITE, K.E., presented Institution of Mechanical Engineers Conf.Measurement and Techniques in Air
      and Water Pollution, London, 1972.
[20] INTERNATIONAL COMMISSION ON RADIOLOGICAL PROTECTION, Report of Committee II on Permissible
      Dose for Internal Radiation (1959), ICRP Publication No.2, Pergamon Press, Oxford (1960).
[21] GOLDBERG, E.C., BROECKER, W.S., GROSS, M.G., TUREKIAN, K.K., in Radioactivity in the Marine
      Environment, National Academy of Sciences (1971) Chap.5.
[22] LOWMAN, F.G., RICE, T.R., RICHARDS, F.A., ibid., Chap.7.
[23] BOWEN, V.T., OLSEN, J.S., OSTERBERG, C.L., RAVERA, J., ibid., Chap.8
[24] DUTTON, J.W.R., Ministry of Agriculture, Fisheries and Food, Lowestoft, Tech.Rep.FRL 4 (1969).
[25] WOOLNER, L.E., to be published.
[26] MITCHELL, N.T., Radioactivity in Surface and Coastal Waters of the British Isles 1967, Ministry of
      Agriculture, Fisheries and Food, Lowestoft, Tech.Rep.FRL 2 (1968).
[27] MITCHELL, N.T., ibid. for 1968, FRL 5 (1969).
[28] MITCHELL, N.T., ibid. for 1969, FRL 7 (1971).
[29] MITCHELL, N.T., ibid. for 1970, FRL 8 (1971).
[30] TSURUGA, H., Bull.Japan Soc.Sci.Fish. 31 (1965) 651-58.
[31] BONI, A.L., Anal.Chem. 32 (1960) 599-604.
[32] WONG, K.M., referred to by BOWEN, V.T., in Reference Methods for Marine Radioactivity Studies,
      Technical Reports Series No.118, IAEA, Vienna (1970) 93-127 (Appendix D).
[33] SAMSAHL, K., Rep. AE-82 (1962).
[34] SUNDERMAN, D.N., TOWNLEY, C.W., Radiochemistry of Silver, NAS-NS 3047 (1961).
[35] HANDLEY, T.H., DEAN, J.A., Anal.Chem. 32 (1960) 1878-83.
[36] HODGE, V.F., FOLSOM, T.R., Anal.Chem. 44 (1972) 381-83.
[37] REYNOLDS, E., Health Phys. 20 (1971) 241-51.
[38] RILEY, J.P., in Chemical Oceanography (RILEY, J.P., SKIRROW, G., Eds), Academic Press, London
      (1965) 295-424.
[39] NODDACK, I., NODDACK, W., Ark.Zool. 32A (1940) 1-35.
[40] LAI, M.G., WEISS. H.V., Anal.Chem. 34 (1962) 1012-15.
[41] BLACK, W.A.P., MITCHELL, R.L., J.Mar.Biol.Assoc.UK 30 (1951) 575-84.
[42] DAVANKOV, A.B., LAUFER, V.M., AZHAZHA, E.G., GORDEIVSKII, A.V., KIRYUSHOV, V.N.,
      Izv.Vyssh.Uchebn.Zaved. 5 (1962) 118.
[43] SOYER, J., Vie et Milieu 14 (1963) 1-36.
[44] SCHUTZ, D.F., TUREKIAN, K.K., Geochim.Cosmochim.Acta 29 (1965) 259-313.
[45] ROBERTSON, D.E., Pacific Northwest Laboratory Ann.Rep. for 1970, BNWL-1551: Pt.2 (1971) 12-15.

[46] BROOKES, R.H., Talanta 12 (1965) 511-16.

[47] PRESTON, A., JEFFERIES, D.F., DUTTON, J.W.R., HARVEY, B.R., STEELE, A.K., Environ.Pollut. 3 (1972) 69-82.

[48] WOODS HOLE OCEANOGRAPHIC INSTITUTE, in Trace Element Intercalibration Study, BREWER, P.G., SPENCER, D.W., unpublished manuscript, 1970.

[49] TUREKIAN, K.K., SCHUTZ, D.F., J.Geophys.Res. 70 (1965) 5519-28.

[50] BROOKES, R.R., PRESLEY, B.J., KAPLAN, I.R., Talanta 14 (1967) 809-16.

[51] JOYNER, T., HEALY, M.L., CHAKRAVATI, D., KOYANAGI, T., Environ.Sci.Technol. 1 (1967) 417-24.

[52] NICHOLLS, G.D., CURL, H., BOWEN, V.T., Limnol.Oceanogr. 4 (1959) 472-78.

[53] TUREKIAN, K.K., SCHUTZ, D.F., quoted by TUREKIAN, SCOTT, Environ.Sci.Technol. 1 (1967) 940-42.

[54] RANCITELLI, L.A., Pacific Northwest Laboratory Ann.Rep. for 1968, BNWL-1051: Pt. 2 (1969) 146-51.

[55] TUREKIAN, K.K., Geochim.Cosmochim.Acta 32 (1968) 603-12.

[56] CHAO, T.T., BALL, J.W., NAKAGAWA, H.M., Anal.Chim.Acta 54 (1971) 77-81.

[57] LUECKE, W., EMMERMANN, R., At.Absorpt.Newslett. 10 (1971) 45-49.

[58] BOWEN, H.J.M., Analyst 92 (1967) 124-31.

[59] GIRARDI, F., PAULY, J., SABBIONI, E., VOS, G., Nuclear Activation Techniques in the Life Sciences (Proc.Symp.Amsterdam, 1967), IAEA, Vienna (1967) 229-46.

[60] NADKARNI, R.A., EHMANN, W.D., J. Radioanal.Chem. 3 (1969) 175-85.

[61] RANCITELLI, L.A., Pacific Northwest Laboratory Ann.Rep. for 1968, BNWL-1051: Pt. 2 (1969) 131-34.

A REVIEW OF METHODS FOR THE DETERMINATION OF IODINE-131 IN THE MARINE ENVIRONMENT

J.W.R. DUTTON
Ministry of Agriculture, Fisheries and Food,
Fisheries Radiobiological Laboratory,
Lowestoft, Suffolk,
United Kingdom

Abstract

A REVIEW OF METHODS FOR THE DETERMINATION OF IODINE-131 IN THE MARINE ENVIRONMENT.
The origins of the radionuclide of iodine — from fall-out and radioactive waste disposal — are summarized, together with its nuclear properties. Methods of analysis for this radionuclide are reviewed, especially those suitable for the control of radioactive waste disposal. In this context direct γ-spectrometric analysis of the dried material is generally suitable, but more complex techniques (including radiochemical separations) may be required for ecological research. A summary of methods for the determination of the stable elements in marine materials is also presented.

1. INTRODUCTION

Iodine-131 is produced in fission with a yield of about 3%; it is a β^- emitter (β_{max} 0.61 MeV) and also emits several γ-photons, the most abundant of which has an energy of 0.36 MeV [1]. Although its has been detected and measured in the marine environment as a result of local fall-out from nuclear weapon testing [2, 3], it is not generally encountered in stratospheric fall-out, due to its short half-life of 8.04 days. For the same reason it is not discharged in detectable quantities as a result of waste disposal from irradiated fuel reprocessing centres. Reactor operations, however, do result in occasional discharges of ^{131}I; for example, it has been measured in coastal seawaters and seaweeds as a result of marine discharges from the nuclear power station at Tarapur in India [4].

Although no ^{131}I has been detected in the marine environment as a result of waste disposal operations from UK nuclear-powered electricity generating stations, reactor development operations at Winfrith have resulted in a pulsed discharge of ^{131}I into the sea [5]. The opportunity was taken to study the movement of the radionuclide in the environment to test models for the assessment of accident situations [6] and to check analytical techniques. The analytical methods presented in this paper are those used at this laboratory during these investigations, in which ^{131}I was measured in seawater and Fucus weed.

2. ANALYSIS FOR IODINE-131 IN THE CONTEXT OF RADIOACTIVE WASTE DISPOSAL

The maximum acceptable levels of ^{131}I in various marine seafoods are shown in Table I. These derived working limits (DWL) are calculated from the data recommended by the International Commission on Radiological

TABLE I. [131]I LEVELS IN SEAFOODS BASED ON ICRP
RECOMMENDATIONS

ICRP public daily maximum intake	4.4×10^3 pCi		
Material	Fish	Seaweed	Shellfish
Consumption per day	1 kg	500 g	100 g
DWL	4.4 pCi/g	8.8 pCi/g	44 pCi/g
1% of DWL	0.04 pCi/g	0.09 pCi/g	0.44 pCi/g

Protection (ICRP) Committee 2 [7] for ^{131}I levels in drinking water and rates
of consumption of marine seafoods — which may, of course, vary in
different localities. 1% of these DWLs are shown as an indication of required
lower limits of detection for control purposes, and it can be seen that the
determination of ^{131}I at levels of about 0.05 pCi/g of the critical material
will be required if sustained discharges to the marine environment are being
considered.

3.   THE DETERMINATION OF IODINE-131 IN SEAWATER

Several methods for the determination of radioiodine in natural waters
have been published (e.g. Refs [8,9]), which rely on carrier equilibration,
by oxidation to iodate and reduction to iodide, followed by conversion to
free iodine and extraction into carbon tetrachloride; the iodine is back-
extracted into the aqueous phase and precipitated as silver iodide for subsequent counting. This method was used as the basis of our separation of
[131]I from seawater, but during our investigations we found that traces of
sodium chloride were carried through the separation stages and precipitated
with the silver iodide; we therefore used Pd^{II} to precipitate palladous
iodide — palladous chloride is soluble [10]. We considered it essential to
carry out the oxidation and reduction stages to ensure complete equilibration
of radioactive and stable forms when measuring ^{131}I in seawater, in which
a significant fraction of the iodine present is in the iodate form.

The method employed is as follows:

(i) Transfer 5 litres of the sample into a large beaker and, after the
addition of 50 mg of iodine carrier as sodium iodide solution, adjust the pH
to about 9 with ammonia solution (SG 0.880).

(ii) Add 30 ml 10% sodium hypochlorite solution and allow to stand for
30 minutes.

(iii) Add 10 g hydroxylamine hydrochloride, followed by 250 ml carbon
tetrachloride and 40 ml nitric acid (SG 1.42). N.B., carbon tetrachloride is
toxic and any operation involving its use should be carried out in a well-
ventilated fume-hood.

(iv) Add 5 g sodium nitrite and extract the liberated iodine into the
carbon tetrachloride using a mechanical stirrer.

(v) Remove the organic phase and back-extract the iodine by shaking with 20 ml saturated aqueous sulphur dioxide solution and 80 ml distilled water in a separating funnel.

(vi) Return the carbon tetrachloride to the large beaker and perform a second extraction after the addition of a further 5 g sodium nitrite.

(vii) Repeat the back-extraction, section (v), and combine the two aqueous extracts.

(viii) Add a few drops of 12N hydrochloric acid and allow to stand for at least one hour, sufficient to enable most of the sulphur dioxide to disperse.

(ix) Add 50 mg Pd^{II} as palladous chloride solution and filter the precipitate of palladous iodide on a tared filter paper.

(x) Dry on a desiccator and weigh to determine the chemical yield.

A thin end-window β^- counter was used to measure the count rates of the sources, which were converted to activity in pCi by comparison with the count rate of a standard source similarly prepared. A counting efficiency of about 45% was obtained (i.e. 1 count/min is equivalent to 1 pCi), and 0.2 pCi ^{131}I/l in the seawater resulted in a count-rate of 1 count/min, equal to that of the background.

4.   THE DETERMINATION OF IODINE-131 IN BIOLOGICAL MATERIALS

Fucus seaweed does not, of course, figure in any critical pathway for the control of radioiodine discharges, but it was decided to determine ^{131}I in this commonly occurring weed in order to measure the dispersion of the radionuclide along the coast, following the pulsed discharge. Preliminary calculations indicated that the use of direct γ-spectrometry, using our standard 75 mm $\times$ 75 mm NaI(Tl) crystal, would be suitable as long as sufficient sample was available. To obviate any uncertainty about loss of ^{131}I on drying, the wet sample was analysed. An annular container 220 mm (diameter) $\times$ 125 mm (depth) was constructed to hold 2 kg of wet weed; this was placed around the crystal and the sample counted for 60 minutes. By measuring the peak height and comparing it with those of standards similarly prepared, levels of iodine down to 0.1 pCi/g were measured. Several of the samples were then oven-dried at 100-110°C and the ^{131}I content determined by our routine γ-spectrometric method [11]; similar values ($\pm$ 15% of 'wet' result) were obtained, with no systematic error, indicating that, within the limits of precision of the method, there was no loss of iodine on drying the sample. No limit of detection has been determined for this method, but levels of ^{131}I of the order of 0.05 pCi/g wet could easily be measured using 250 g of the dried sample.

5.   COMMENTS ON THE METHODS

For biological materials direct γ-spectrometric analysis of wet or oven-dried material is satisfactory, producing data with adequate precision down to ^{131}I levels of 0.1 pCi/g or less.

For seawater a simpler method should be used if the requirement for ^{131}I analysis occurs again. The present method, based largely on standard practice, is not too time-consuming but it is lengthy.

REFERENCES

[1] THE RADIOCHEMICAL CENTRE, Amersham, The Radiochemical Manual, 2nd ed. (1966) 327 pp.

[2] DONALDSON, L.R., SEYMOUR, A.H., HELD, E.E., HINES, N.O., LOWMAN, F.G., OLSON, P.R., WELANDER, A.D., USAEC Rep. UWFL-46 (1956).

[3] PALUMBO, R.F., USAEC Rep. UWFL-87 (1963).

[4] KAMATH, P.R., IYENGAR, M.A.R., BHAT, I.S., in Symp. Natural Radiation Environment II, Houston, 1972.

[5] MITCHELL, N.T., Radioactivity in Surface and Coastal Waters of the British Isles 1968, Ministry of Agriculture, Fisheries and Food, Lowestoft, Tech. Rep. FRL.5 (1969).

[6] PRESTON, A., JEFFERIES, D.F., Environmental Contamination by Radioactive Materials (Proc. Seminar Vienna, 1969), IAEA, Vienna (1969) 183-211.

[7] INTERNATIONAL COMMISSION ON RADIOLOGICAL PROTECTION, Report of Committee II on Permissible Dose for Internal Radiation (1959), ICRP Publication No.2, Pergamon Press, Oxford (1960).

[8] WORLD HEALTH ORGANIZATION, Methods of Radiochemical Analysis (1966) 163 pp.

[9] UNITED KINGDOM ATOMIC ENERGY AUTHORITY, P.G. Rep. 204 (1961).

[10] KLEINBERG, J., COWAN, G.A., The Radiochemistry of Fluorine, Chlorine, Bromine and Iodine, NAS-NS-3005 (1960).

[11] DUTTON, J.W.R., Ministry of Agriculture, Fisheries and Food, Lowestoft, Tech. Rep. FRL.4 (1969).

A RAPID METHOD FOR THE DETERMINATION OF RADIOIODINE IN SUSPENDED MATTER AND SEAWATER

S. WLODEK
Central Laboratory of Radiological Protection,
Warsaw, Poland

Short Communication

As contamination of coastal waters by accidental releases of fission products from nuclear reactors, sited more and more frequently in coastal areas, becomes increasingly possible, one of the most dangerous radio-nuclides will be [131]I. Although it has only a short half-life, this radionuclide could be a grave health hazard to inhabitants and tourists using the beaches and bathing in the sea, and for those people who consume various marine organisms such as seaweed (in laverbread), oysters, anthropodes and fish etc.

It is clear that a rapid method is essential for measuring the concentration of iodine-131 in seawater and eventually in all suspended matter (Seston).

The classic gravimetric method of Fairman [1] for the measurement of iodine as a precipitate of silver iodide has its difficulties: the analysis takes a long time, the silver chloride dissociates under the influence of light, and the chemical yield is on the whole poor. The Central Laboratory of Radiological Protection, Warsaw, is in the process of testing a sensitive and rapid method for the measurement of radioiodine in seawater. The method is based on well-documented procedures. The initial stage is to separate the iodine from other components in the sample by solvent-extraction using carbon tetrachloride, as used by Sansoni [2]. Next the iodine is collected on an ion-exchange resin. The novelty of this procedure is that the form of this resin permits direct measurement of the radioiodine with high sensitivity using β-spectrometry with anticoincidence shielding. The analysis is rapid and takes about 1 hour.

The method measures the inorganic form of iodine. From experiments that have been performed it has been shown that inorganic iodine in rainwater, continental surface waters and in water of the Baltic Sea can readily be measured by this method.

It is expected that the present work will be expanded to look at the concentration of iodine in suspended matter in seawater. Theoretically, iodine in suspended material can be easily solubilized by treatment with sodium hydroxide or by fusion with sodium carbonate [3]. This would, however, cause an increase in the time required for the analysis.

It is hoped that next year the measurement of iodine in suspended matter of the Baltic Sea will begin and the method and results will be published in due course.

REFERENCES

[1] FAIRMAN, D., Rep. HNL-6887, Chemistry TID-4500.
[2] SANSONI, B., Angew. Chem. 73 (1961) 493.
[3] MORGAN, A., MITCHELL, G.R., A New Method for the Determination of Iodine-131 in Herbage Including Measurements Made in Autumn 1961, Rep. AERE-M 1004.

A METHOD FOR THE RAPID ENRICHMENT AND DETERMINATION OF SILVER-110m AND IODINE-131 FROM SEAWATER

H.-F. EICKE
Deutsches Hydrographisches Institut,
Hamburg,
Federal Republic of Germany

Abstract

A METHOD FOR THE RAPID ENRICHMENT AND DETERMINATION OF SILVER-110 m AND IODINE-131 FROM SEAWATER.

The procedure discussed employs isotopic exchange of solid, inactive silver iodide (AgI) for the separation and enrichment of radioactive iodine isotopes and silver isotopes from seawater. The water to be investigated is siphoned through a layer of silver iodide. To increase the exchange capacity of the silver iodide surface, it is precipitated onto asbestos threads (Gooch asbestos). A detailed working procedure for the analysis is described. The influence of interference from other fission products is discussed. Under the conditions described about 90% of the $^{110}Ag^m$ can be fixed by the exchanger and values of 70-75% have been obtained for ^{131}I. Up to 25 litres of seawater can be processed. The time required for analysis and measurement and the limits of detection are discussed.

INTRODUCTION

Direct γ-spectrometry of radioactive iodine and silver in seawater is only partially possible as their lines in the spectrum can be distorted or masked by other radionuclides present in the mixture and because the activities are usually not sufficient for direct determination. Scanning the literature for suitable methods for the determination of silver-110m and iodine-131 in seawater suggested that ion and isotope exchange processes might be the most successful [1-6]. Jaworowski's method [2] (also described in IAEA Technical Reports Series No. 95 [7]) for the determination of radioactive iodine in urine and rainwater was investigated for its suitability for the determination of ^{131}I in seawater.

In the course of these experiments it seemed logical to investigate the exchange behaviour of silver-110m on solid silver iodide, the more so since analogous investigations into the exchange of silver on solid silver chloride had already shown good results [8, 9].

Jaworowski's method measures the β-radiation with a Geiger-Müller tube. We decided to use γ-spectrometrical measurement since, during experiments with seawater, a partial separation of zirconium-95 (niobium-95) on silver iodide had been observed.

PRINCIPLES OF THE EXCHANGE PROCEDURE

Corresponding to the mechanism of simple heterogeneous isotope exchange, the seawater to be investigated is slowly drawn off over solid silver iodide, which, to increase the exchange capacity of the surface, is deposited on asbestos fibres.

179

TABLE I.   RELATIONSHIP BETWEEN CURRENT VELOCITY
THROUGH THE COLUMN AND YIELD OF $^{110}Ag^m$

Experiment No.	Current velocity (ml/min)	Count rate (counts/min)	Yield (%)
6	170	102	89
7	200	103	90
8	200	107	93
9	250	107	93
10	250	103	90

The calibration sample produced 115 counts/min $\pm$ 1.5%.

EXPERIMENTAL DETAILS

Equipment

Siphoning equipment with water jet injector pump;
Aspirator and Büchner funnel, ϕ 7 cm (funnel with a fixed, latticed
   glass plate);
Filter paper, ϕ 7 cm;
Measuring cylinders, 500 and 100 ml;
Petri dish, ϕ 7.2 cm;
Sodium iodide (Tl) crystal, 60 $\times$ 60 mm;
Multi-channel analyser with additional equipment for evaluation.

Methods

To prepare the exchanger suspension the following substances were
added in succession to 3 litres of distilled water:

10 ml	conc. H_2SO_4
3 g	Sodium iodide (NaI) dissolved in 50 ml 1$\underline{N}$ H_2SO_4
50 g	Gooch asbestos
7.5 g	Silver nitrate ($AgNO_3$) dissolved in 10 ml water

The suspension must be stored in darkness. These amounts are sufficient
for 10 analyses.
To carry out the exchange, 300 ml of well-shaken suspension was
placed in several portions in the Büchner funnel, which had been lined
with filter paper. This was then drawn off with a gentle vacuum, the
moist filter contents covered with a paper filter and washed twice with
50 ml 1$\underline{N}$ H_2SO_4. 4 ml of H_2SO_4 was added to each litre of the seawater
to be investigated, and this was sucked through the exchanger with a

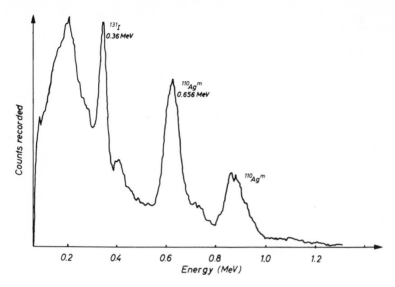

FIG.1. Spectrum of a solution of $^{110}Ag^m$ and ^{131}I.

current velocity of 200-250 ml/min. This was then washed twice with
50 ml 0.1N H$_2$SO$_4$ and sharply siphoned. The exchanger was placed in a
Petri dish and measured with a sodium iodide crystal. The manufacture
of the calibration preparation can follow quite simply. The exchanger,
as prepared in the analysis, is as far as possible equally prepared with
a known $^{110}Ag^m$ activity and measured by the same geometry as that of the
analysis samples.

DETERMINATION OF RADIOCHEMICAL YIELD FROM $^{110}Ag^m$

Unfiltered North Sea water from 57°N 8°E with a salinity of about
32‰ was used for all experiments.

Preliminary experiments to decide upon an ideal current speed
through the exchanger with the aim of obtaining the highest yield of
$^{110}Ag^m$ resulted in the values given in Table I. These refer to 1-litre
samples, to each of which was added 1000 pCi $^{110}Ag^m$.

Extension of these experiments with larger amounts of water revealed
that with throughputs of up to 20 litres 90% of the silver activity was still
fixed on the exchanger. The current velocity of 250 ml/min gradually
decreased slightly during the run-through of the 20-litre samples, so that
2 hours were required to siphon the whole amount through the exchanger.

The silver activity was also varied in these experiments. As an
example, additions of 100 pCi $^{110}Ag^m$ to each 20 litres of seawater (equivalent to 5 pCi/l) with a measurement period of 40 minutes still produced
impulse rates whose statistical error lay below 10%. Using 20 litres of
seawater for isotope exchange, the limit of detection for $^{110}Ag^m$ was
2 pCi/l (± 20% statistical error).

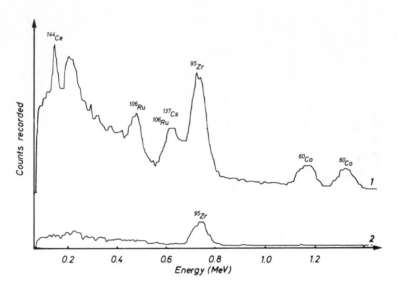

FIG.2. (1) Spectrum of interfering radionuclides before passing through exchanger; (2) spectrum of exchanger after passage of nuclides.

In experiments employing even larger quantities of water (up to 35 litres) the yield of Ag activity on the exchanger was reduced. Therefore not more than 25 litres of water should be used where possible.

INFLUENCE OF OTHER RADIONUCLIDES ON THE DETERMINATION OF $^{110}Ag^{m}$

When isotope exchange reactions are also very selective one must devote particular attention to the question of possible disturbance from other nuclides, which may possibly remain on the exchanger owing to ion reactions, build-up of complexes, or electrostatic and adsorptive energy. Therefore, the exchanger behaviour of artificial radionuclides, the presence of which is to be expected in seawater, must be examined. Appropriate experiments, a brief account of which follows, were carried out on $^{144}Ce/^{144}Pr$; $^{106}Ru/^{106}Rh$; $^{137}Cs/^{137}Ba$; $^{95}Zr/^{95}Nb$; ^{60}Co and ^{54}Mn. As clarification of these experiments several γ-spectra are illustrated in Figs 1-4.

Figure 1 shows the energy state of the peaks that were used for γ-spectrometric determination of $^{110}Ag^{m}$ (0.656 MeV) and ^{131}I (0.36 MeV).

To ascertain the influence of these nuclides upon the exchanger, a mixture of

880 pCi	$^{144}Ce/^{144}Pr$
2400 pCi	$^{106}Ru/^{106}Rh$
1100 pCi	$^{137}Cs/^{137}Ba$
1100 pCi	^{95}Zr
2200 pCi	^{95}Nb
1300 pCi	^{60}Co

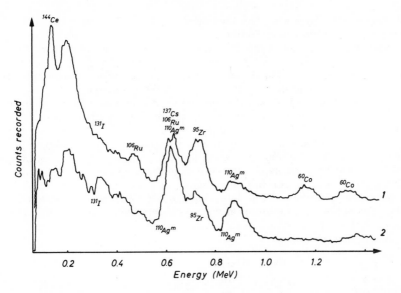

FIG.3. (1) Spectrum of mixture of silver and iodine nuclides in seawater before passing through exchanger; (2) spectrum of exchanger after passage of nuclides.

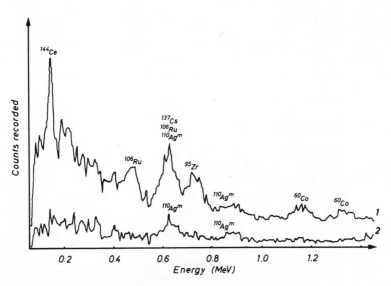

FIG.4. (1) Spectrum of nuclide mixture with the addition of 100 pCi $^{110}Ag^m$; (2) Spectrum of exchanger after passage of the mixture diluted with 20 litres of seawater.

was placed in a Petri dish and measured, using the same measuring
geometry as for the silver iodide exchanger. The spectrum obtained is
reproduced in Fig.2 as curve 1. The nuclide mixture, diluted with 1 litre
of acidified seawater, was siphoned through the exchanger. The exchanger
measurement provided curve 2 in Fig.2. One can see quite clearly that
only a certain amount of Zr/Nb remains on the exchanger.

Figure 3, curve 1 shows the spectrum of a measurement taken in a
ring pot device. The nuclide mixture, diluted with 1 litre of seawater,
contained 500 pCi ^{110}Ag^m and 1100 pCi ^{131}I, as well as the afore-mentioned
activities. After this solution had been run through the exchanger the
measurement shown in curve 2 was obtained. Comparison of both spectra
clearly shows the good separation effect of the exchanger. The ^{95}Zr remaining on the exchanger in the presence of very high Zr/Nb activities
can cause interference in the determination of ^{110}Ag^m. The physical adsorption of zirconium and niobium on silver chloride and silver iodide has also
been observed by other authors [6, 9].

A mixture of:

300 pCi	^{144}Ce/^{144}Pr
350 pCi	^{106}Ru/^{106}Rh
200 pCi	^{137}Cs/^{137}Ba
100 pCi	^{95}Zr
200 pCi	^{95}Nb
280 pCi	^{60}Co
100 pCi	^{110}Ag^m

was placed in a Petri dish and the γ-spectrum taken (Fig.4, curve 1).
After dilution of the nuclide mixture with 20 litres of acidified seawater,
it was passed through the exchanger and the exchanger measured (Fig.4,
curve 2). The concentration of 5 pCi ^{110}Ag^m per litre present in this
experiment is clearly seen in the exchanger spectrum. An impulse measurement of 9.6 counts/min for a period of 40 minutes was obtained. If one takes
as a basis the 90% yield given by earlier investigations, one obtains a total
activity of 92 pCi on the exchanger, which is equivalent to 4.6 pCi/1 of ^{110}Ag^m.
Zirconium and niobium do not appear on the exchanger in the range of
activity used.

INVESTIGATIONS WITH IODINE-131

The determination of ^{110}Ag^m on the silver iodide exchanger provided
an opportunity of testing the application of the methods used by Jaworowski [2]
and Jacobs and Lehmann [6] for the detection of iodine-131 in seawater. We
obtained different yields compared with Jaworowski's results when we ran
1 litre of seawater with 1700 pCi ^{131}I added. From 1 litre of seawater we
obtained a separation of 70-75% of the ^{131}I added (Jaworowski, 97% in rainwater). After increasing the quantity of seawater put through the exchanger
to 5 litres, the amount precipitated was reduced to about 65%. The precipitation obtained would be even more severely reduced by the use of larger
quantities of water. We do not know the reason for this. An improvement
in the precipitation rate to 80% can be obtained by the addition of 10 ml of
a 1% solution of sodium thiosulphate to 1 litre of seawater (acidified with

4 ml conc. H_2SO_4). This illustrates the dependence of iodine precipitation on the chemical condition.

The results show that the present method for the determination of iodine-131 in seawater may be carried out without hesitation with quantities of water up to 1 litre. The use of larger quantities will produce inaccurate results.

RESULTS

Isotope exchange on AgI permits a relatively rapid concentration and specific selection of $^{110}Ag^m$ from seawater. The time required depends on the quantity of water used for the exchange. The run-through speed was about 200-250 ml/min. The limit of detection was 2 pCi/l using at least 20 litres of seawater. Under these conditions a measuring period of at least 40 minutes is necessary. The quantity of water and the measuring time required decrease with higher activities.

Certain disturbances can be caused owing to the presence of large quantities of ^{95}Zr; however, these can be eliminated by taking γ-spectra.

REFERENCES

[1] SUNDERMAN, D.N., TOWNLEY, C.W., The Radiochemistry of Silver, Nuclear Science Series NAS-NS 3047 (1961).

[2] JAWOROWSKI, Z., Determination of iodine in urine and rainwater, Nukleonika (Warsaw) 5 (1960) 81-86.

[3] OVERMAN-CLARK, Radioisotope Techniques, McGraw Hill, New York (1960).

[4] HAINBERGER, L., SANCHEZ, S., Separation of isotopes, Mikr. Acta 1 (1962) 87-91.

[5] HORSHI, J., Isotope Exchange of Radioiodine, INR-636/XIII/J, Warsaw (1965) 83.

[6] JACOBS, H., LEHMANN, D., Schnellbestimmung von Radiojod in natürlichen Wässern, Rep. KFA Jülich 410 (1966).

[7] INTERNATIONAL ATOMIC ENERGY AGENCY, Quick Methods for Radiochemical Analysis, Technical Reports Series No.95, IAEA, Vienna (1969) 15.

[8] SUNDERMAN, D.N., MEINKE, W.W., Evaluation of radiochemical separation procedures, Anal. Chem. 29 (1957) 1578.

[9] KAMATH, P.R., Assessment of Radioactivity in Man (Proc. Symp. Heidelberg, 1964), IAEA, Vienna (1964) 195-215.

A STUDY OF THE STABILITY OF
SOME RADIONUCLIDES IN SEAWATER

J. W. R. DUTTON, R. D. IBBETT, L. E. WOOLNER
Ministry of Agriculture, Fisheries and Food,
Fisheries Radiobiological Laboratory,
Lowestoft, Suffolk,
United Kingdom

Abstract

A STUDY OF THE STABILITY OF SOME RADIONUCLIDES IN SEAWATER.

An investigation has been made into the storage stability in filtered seawater of ruthenium-106, cerium-144 and zirconium-95, present as a result of discharges from the British Nuclear Fuels Ltd. processing plant at Windscale on the Cumberland coast. Both ruthenium-106 and zirconium-95 are stable at pH levels of 1.8 to about 8 (the natural pH of seawater), but cerium-144 is not.

INTRODUCTION

Disposal of low-level liquid radioactive waste [1] to the marine environment of the UK is permitted under carefully controlled conditions; as a result, a number of products can be identified, such as radionuclides of cerium (^{141}Ce and ^{144}Ce), ruthenium (^{106}Ru and ^{103}Ru), caesium (^{134}Cs and ^{137}Cs), strontium (^{89}Sr and ^{90}Sr), and zirconium (^{95}Zr). Although the analysis of seawater is not usually necessary for radiological control monitoring purposes [2] — analysis of the critical radionuclide in the appropriate material is usually sufficient — the determination of radioactivity in the seawater is necessary in connection with monitoring applied to research studies such as the rate of dispersion of the different radionuclides, their rate of removal by sediments, and the determination of concentration factors.

The analysis of seawater for radioactivity can pose difficult problems, not the least being the stability of the radionuclides during storage; the delay between sampling (often on board ship) and chemical processing at the laboratory is often lengthy, and the position is aggravated by the large number of samples that are collected at one time, especially on sea cruises. It could be argued that chemical processing, or at least an initial separation on board ship, is the answer to these problems, but the number of scientists that can be accommodated on board is limited, and the emphasis is therefore placed on the collection of samples rather than on their analysis. Simple techniques such as the passage of seawater through ion exchange columns are certainly feasible — for example the use of potassium cobaltihexacyanoferrate columns to remove radionuclides of caesium — but if there is uncertainty about the speciation of the nuclides, then carrier addition and isotopic equilibration (often resulting in too complex a procedure) must be carried out before any initial separation.

We therefore undertook an investigation into the storage stability of a group of radionuclides that were of special interest to the laboratory and about which little or no information on stability existed. On the basis of discharge data, half-life and also, to an extent, relative ease of analysis,

TABLE I. RUTHENIUM-106 VALUES: pCi/l CORRECTED TO DAY 0

Container type	pH level														
	Day 7			Day 14			Day 28			Day 91			Day 210		
	1.8	3.0	$\underline{N}$	1.8	3.0	$\underline{N}$	1.8	3.0	$\underline{N}$	1.8	3.0	$\underline{N}$	1.8	3.0	$\underline{N}$
A	61.5	64.2	73.2	72.4	65.3	(22.3)[a]	64.3	71.5	74.7	71.6	75.1	66.3	73.4	74.2	71.6
B	67.6	69.9	69.8	71.0	70.1	68.5	75.3	66.6	61.6	71.6	68.5	70.4	69.7	70.3	72.2
C	70.9	74.8	73.9	62.9	68.7	65.2	69.2	68.7	65.8	68.8	66.5	73.4	65.4	71.1	64.9
D	64.8	68.7	70.6	70.3	67.9	64.0	69.5	66.8	69.5	71.8	67.2	71.5	65.7	70.3	73.1
E	(48.5)[b]	72.1	75.6	66.0	72.5	65.7	65.6	61.9	71.8	75.6	67.5	71.4	69.1	69.9	65.5
F	66.4	71.9	68.7	63.4	73.1	69.2	65.9	68.3	65.7	71.1	71.6	71.1	68.5	72.0	67.7

[a] Rejected by Chauvenet's criterion and replaced by 68.3 using least-squares fit technique.
[b] Rejected by Chauvenet's criterion and replaced by 64.7 using least-squares fit technique.

we chose ^{95}Zr, ^{106}Ru and ^{144}Ce for the investigation. It was decided to omit ^{90}Sr because it is present with a large amount of natural carrier (8 mg/l), presumably in equilibrium, and there existed some previously reported data [3,4] on the stability of strontium in seawater. For similar reasons the storage stability of ^{137}Cs was also not investigated.

There was never any question of our using radioactive tracer additions. We have often found differences between the chemical reactivity of a radio-nuclide purchased from the Radiochemical Centre Ltd., Amersham, and that of the same radionuclides found in the environment; these variations can only be ascribed to differences either in the species or in the radio-nuclide to stable carrier ratios, which, in the case of cerium, zirconium and ruthenium, are difficult to match with those in seawater. The obvious material to use in these investigations was the seawater from the vicinity of a radioactive waste discharge point, which would contain these radio-nuclides in the relevant physical and chemical states, whatever they might be.

The proliferation of types of plastic materials, coupled with the unfortunate trend to use trade-names, makes the choice of containers for the storage of seawater a difficult one. Polyethylene (both high- and low-density, or linear and cross-linked), polypropylene and polyvinylchloride (PVC) are three commonly used materials, because of their physical characteristics and relative cheapness, and containers made of these materials were used in this investigation.

Several methods may be employed to ensure the stable storage of seawater, including freezing, acidification, the addition of oxidizing, reducing or complexing agents, and the addition of carriers. Freezing was impracticable because of the relatively large size of sample (≮ 25 litres); carrier addition was also ruled out for this first investigation because of the care that needs to be taken over the addition if chemical yields had later to be determined. The addition of other chemical reagents (oxidants, reductants, complexants, etc.) was rejected because of lack of knowledge of their effects in seawater, and the difficulty of selecting reagents that were applicable to all the radionuclides. Our choice, there-fore, was to assess the value of acid addition, and we decided upon hydro-chloric acid, since Cl$^-$ is already the major anion in seawater, and the use of this acid would least affect the subsequent chemical processing.

EXPERIMENTAL

Seawater was collected from the vicinity of the discharge point of the British Nuclear Fuels Ltd. factory at Windscale on the Cumberland coast in November 1969 during a research cruise on RV CORELLA. 2500 litres were pumped aboard into a polyethylene tank, using a plastic impeller pump and PVC hose strengthened with nylon. The water was recirculated, using the pump and a header tank, for 3 hours and then filtered through 0.22 μm pore-size Millipore filter papers [5]; the pressurized filtering apparatus passed the filtered seawater into a second polyethylene tank and, on completion of the filtering, the water was again recirculated to ensure complete mixing. At the end of this period the water was divided between six groups of containers: high-density polyethylene (two types: US manu-facturer (type A) and UK manufacturer (type B)), low-density polyethylene (type C), polypropylene (type D) and PVC (two types: US manufacturer

TABLE II. RUTHENIUM-106 VALUES: CONTAINER versus DAY, Σ pH ACTIVITIES i.e. pCi/3 l

Container type	Day 7	Day 14	Day 28	Day 91	Day 210	Σx_{ij}^2	$\dfrac{(\Sigma x_{ij})^2}{5}$	$\bar{x}_{ij}$
A	198.9	206.0	210.5	213.0	219.2			209.5
B	207.3	209.6	203.5	210.5	212.2			208.6
C	219.6	196.8	203.7	208.7	201.4			206.0
D	204.1	202.2	205.8	210.5	209.1			206.3
E	207.0	205.7	199.8	214.5	204.5			206.3
F	212.4	204.2	199.3	213.8	208.2			207.6
Σx_{ji}^2						1 291 377	1 290 492	
$\dfrac{(\Sigma x_{ji})^2}{6}$						1 290 727		
$\bar{x}_{ji}$	208.2	204.1	203.8	211.8	209.1			
pCi/l	69.4	68.0	67.9	70.6	69.7			

TABLE III. RUTHENIUM-106 VALUES: DAY versus pH, Σ CONTAINER ACTIVITIES, i.e. pCi/6 l

Day	pH level 1.8	pH level 3.0	N	Σx_{ij}^2	$\dfrac{(\Sigma x_{ij})^2}{3}$	$\bar{x}_{ij}$
7	395.9	421.6	431.8			416.4
14	406.0	417.6	400.0			408.2
28	409.8	403.8	409.0			407.5
91	430.5	416.4	424.1			423.7
210	411.8	427.8	415.0			418.2
Σx_{ji}^2				2 582 550	2 581 453	
$\dfrac{(\Sigma x_{ji})^2}{5}$				2 581 010		
$\bar{x}_{ji}$	410.8	417.4	416.2			
pCi/l	68.5	69.6	69.4			

(type E) and UK manufacturer (type F)). Within these six groups there was a further subdivision based on the pH of storage: at the natural pH of sea-water, and also at pH 3.0 and 1.8, which were obtained by the addition of 0.5 and 2.0 ml 12$\underline{M}$ hydrochloric acid per litre of seawater sample.

After 7, 14, $\overline{28}$, 91 and 210 days' storage containers were opened and the solutions analysed for ^{106}Ru, ^{95}Zr and ^{144}Ce by the methods detailed in Appendix I. In addition, several containers were carefully shredded and gently ignited, and the activity associated with the container itself was also determined; the combination of this activity with that in the seawater from the container gave the activity in the original sample at the date of collection.

The variability of the three criteria of classification — pH level, type of container, and length of storage — was examined using the analysis of variance technique for each of the three radionuclides.

RESULTS AND DISCUSSION

Ruthenium-106

The data obtained for ^{106}Ru are shown in Table I. Using these original data, there was a significant difference between the Day 7 and 14 results and the others. However, it was found that two results on these days were rejected by Chauvenet's criterion [6] and they were replaced by values calculated using a least-squares fit technique [7]. The data were then reassessed by analysis of variance, and the workings are shown in Tables II, III and IV, with the final analysis of variance shown in Table V.

Comparison of the estimated variances of variables and interactions in Table V with the residual variance shows that no significant difference

Text continued on p.196.

TABLE IV. RUTHENIUM-106 VALUES: pH versus CONTAINER, Σ DAY ACTIVITY, i.e. pCi/5 1

pH level	Container type						Σx_{ij}^2	$\dfrac{(\Sigma x_{ij})^2}{6}$	$\bar{x}_{ij}$
	A	B	C	D	E	F			
1.8	343.2	355.2	337.2	342.1	340.4	335.9			342.3
3.0	350.3	345.4	349.8	340.9	350.7	350.1			347.9
$\underline{N}$	354.1	342.5	343.2	348.7	340.4	351.9			348.1
Σx_{ji}^2							2 151 310	2 150 840	
$\dfrac{(\Sigma x_{ji})^2}{3}$							2 150 820		
$\bar{x}_{ji}$	349.2	347.7	343.4	343.9	343.8	346.0			
pCi/l	69.8	69.5	68.7	68.8	68.8	69.2			

TABLE V. RUTHENIUM-106 VALUES: ANALYSIS OF VARIANCE

Source	s.s.	d.f.	Mean square	Components of variance	F ratio
Between containers	19.0	5	3.80	$15\sigma_C^2 + 5\sigma_{CP}^2 + 3\sigma_{CD}^2 + \sigma_0^2$	
pH levels	23.2	2	11.60	$30\sigma_P^2 + 5\sigma_{CP}^2 + 6\sigma_{PD}^2 + \sigma_0^2$	
days	97.2	4	24.3	$18\sigma_D^2 + 3\sigma_{CD}^2 + 6\sigma_{PD}^2 + \sigma_0^2$	1.22
Interactions					
Days versus pH	159.5	8	19.94	$6\sigma_{PD}^2 + \sigma_0^2$	1.93
Containers versus days	197.7	20	9.88	$3\sigma_{CD}^2 + \sigma_0^2$	
pH versus containers	75.0	10	7.50	$5\sigma_{CP}^2 + \sigma_0^2$	
Residual	479.4	40	11.98	σ_0^2	
Total	1 051.0	89			

TABLE VI. CERIUM-144 VALUES: pCi/l CORRECTED TO DAY 0

Container type	Day 7			Day 14			Day 28			Day 91			Day 210		
	1.8	3.0	N̄	1.8	3.0	N̄	1.8	3.0	N̄	1.8	3.0	N̄	1.8	3.0	N̄
A	7.62	7.87	6.49	6.39	6.34	5.60	6.10	4.54	4.14	4.43	3.90	1.24	4.18	3.78	1.32
B	7.53	7.18	6.71	6.02	6.22	5.42	4.61	5.02	4.24	3.71	3.67	2.49	3.38	2.23	2.13
C	7.37	6.45	5.95	6.03	5.87	5.06	4.92	4.99	4.37	3.93	3.58	2.28	3.87	3.51	0.87
D	6.98	7.46	5.72	5.89	5.52	5.49	4.96	4.58	4.61	4.32	2.62	1.84	4.18	2.60	1.70
E	6.99	6.20	5.36	5.83	5.38	4.71	5.02	4.60	4.35	3.59	3.54	1.96	3.80	3.66	1.44
F	6.77	6.07	4.99	5.43	5.04	4.41	4.76	4.67	4.11	3.49	4.04	1.32	3.24	3.84	0.94

TABLE VII. CERIUM-144 VALUES: CONTAINER versus DAY, Σ pH ACTIVITIES, i.e. pCi/3 1

Container type	Day 7	Day 14	Day 28	Day 91	Day 210	Σx_{ij}^2	$\dfrac{(\Sigma x_{ij})^2}{5}$	$\bar{x}_{ij}$
A	21.98	18.33	14.78	9.57	9.28			14.79
B	21.42	17.66	13.87	9.87	7.74			14.11
C	19.77	16.96	14.28	9.79	8.25			13.81
D	20.16	16.90	14.15	8.78	8.48			13.69
E	18.55	15.92	13.97	9.09	8.90			13.29
F	17.83	14.88	13.54	8.85	8.02			12.62
Σx_{ji}^2						6 243	5 600	
$\dfrac{(\Sigma x_{ji})^2}{6}$						6 219		
$\bar{x}_{ji}$	19.95	16.78	14.10	9.32	8.44			
pCi/l	6.65	5.59	4.70	3.11	2.82			

TABLE VIII. CERIUM-144 VALUES: DAY versus pH, Σ CONTAINER ACTIVITIES, i.e. pCi/6 l

Day	pH level		N	Σx_{ij}^2	$\dfrac{(\Sigma x_{ij})^2}{3}$	$\bar{x}_{ij}$
	1.8	3.0				
7	43.26	41.23	35.22			39.90
14	35.59	34.37	30.69			33.55
28	30.37	28.40	25.82			28.20
91	23.47	21.35	11.13			18.65
210	22.65	19.62	8.40			16.89
Σx_{ji}^2				12 696	12 438	
$\dfrac{(\Sigma x_{ji})^2}{5}$				11 505		
$\bar{x}_{ji}$	31.07	28.99	22.25			
pCi/l	5.18	4.83	3.71			

exists between the different containers or between the pH levels; the same is true for the interactions, pH versus containers and containers versus days. Calculation of a new residual variance is therefore possible:

$$\text{New residual variance} = \frac{19.0 + 23.2 + 197.7 + 75.0 + 479.4}{5 + 2 + 20 + 10 + 40}$$

$$= \frac{794.3}{77}$$

$$= 10.32 \text{ with 77 degrees of freedom}$$

Comparison of the interaction of days versus pH with this new estimate of residual variance results in a F ratio of 1.93, so there is no significance in this interaction. Using the variance of 19.94, and substituting this in the components of variance for the 'between days' source, an F ratio of 1.22 is obtained, once again indicating no significant difference in activity levels. There is thus no significant difference in the levels of [106]Ru activity due to variations in storage time between 7 and 210 days, pH levels, and types of containers tested.

The measured Day 0 values for [106]Ru have not been included in this analysis of variance: the data were of necessity obtained by a different analytical technique, involving the shredding of bottles and careful ignition, and, in addition, fewer containers were analysed; this resulted in a different weighting of these data. The Day 0 values obtained were 70.9, 75.9 and 79.1, with a mean value of 75.3 pCi/l, which can be compared with the overall mean of 69.2 (σ on one result ±3.2). A t-test, comparing the

Text continued on p.199.

TABLE IX. CERIUM-144 VALUES: pH versus CONTAINER, Σ DAY ACTIVITIES, i.e. pCi/5 l

pH level	Container type A	B	C	D	E	F	Σx_{ij}^2	$\dfrac{(\Sigma x_{ij})^2}{6}$	$\bar{x}_{ij}$
1.8	28.72	25.25	26.12	26.33	25.23	23.69			25.89
3.0	26.43	24.32	24.40	22.78	23.38	23.66			24.16
N	18.79	20.99	18.53	19.36	17.82	15.77			18.54
Σx_{ji}^2							9 624	9 588	
$\dfrac{(\Sigma x_{ji})^2}{3}$							9 433		
$\bar{x}_{ji}$	24.65	23.52	23.02	22.82	22.14	21.04			
pCi/l	4.93	4.70	4.60	4.56	4.43	4.21			

TABLE X. CERIUM-144 VALUES: ANALYSIS OF VARIANCE

Source	s.s.	d.f.	Mean square	Components of variance	F ratio
Between containers	4.491	5	0.8982	$15\sigma_C^2 + 5\sigma_{CP}^2 + 3\sigma_{CD}^2 + \sigma_0^2$	4.78
pH levels	35.413	2	17.71	$30\sigma_P^2 + 5\sigma_{CP}^2 + 6\sigma_{PD}^2 + \sigma_0^2$	18.6
days	190.91	4	47.73	$18\sigma_D^2 + 3\sigma_{CD}^2 + 6\sigma_{PD}^2 + \sigma_0^2$	50.1
Interactions					
Days versus pH	7.615	8	0.9519	$6\sigma_{PD}^2 + \sigma_0^2$	5.06
Container versus day	3.521	20	0.1761	$3\sigma_{CD}^2 + \sigma_0^2$	1.04
pH versus container	2.882	10	0.2882	$5\sigma_{CP}^2 + \sigma_0^2$	1.71
Residual	6.758	40	0.1689	σ_0^2	
Total	251.59	89			

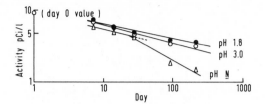

FIG. 1.   Effect of pH on the storage of ^{144}Ce.

TABLE XI.   MEAN ^{144}Ce ACTIVITY AND SURFACE AREA OF CONTAINERS

Mean activity (pCi/l)	Surface area (cm^2/l)	Container
4.93	270	A
4.70	300	B
4.60	300	C
4.56	320	D
4.43	500	E
4.21	720	F

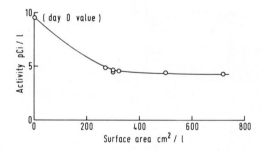

FIG. 2.   The relation between mean ^{144}Ce activity and surface area of container.

two means, resulted in a probability of $0.05 < P < 0.10$, i.e. the means cannot be said to be different.

Over the 210 days of storage, the activity did not drop by $2\sqrt{2 \times 3.2^2}$ (95% confidence) = 9.0 pCi/l, i.e. ≯ an average of 0.06% per day.

Cerium-144

The data obtained for ^{144}Ce are shown in Table VI. The groups of data used for the analysis of variance are shown in Tables VII, VIII and IX with the final calculations of variance in Table X. The estimates of variances of variables (and interactions) are shown in this last table, and

Text continued on p.202.

TABLE XII. ZIRCONIUM-95 VALUES: pCi/1 CORRECTED TO DAY 0

Container level	Day 7			Day 14			Day 28			Day 91			Day 210		
	1.8	3.0	N	1.8	3.0	N	1.8	3.0	N	1.8	3.0	N	1.8	3.0	N
A	11.0	8.6	12.7	12.2	12.1	12.0	11.1	13.8	12.5	11.0	10.8	10.9	10.9	15.0	9.9
B	13.2	12.6	10.9	12.0	11.6	12.2	10.8	12.5	12.6	11.1	10.9	11.3	12.9	16.3	11.2
C	11.1	11.8	12.4	11.9	11.2	12.0	11.9	12.3	12.2	13.1	12.2	13.1	13.0	14.3	11.7
D	10.8	11.6	12.8	11.1	11.5	12.4	11.5	11.7	13.0	13.2	13.2	10.9	14.8	16.1	10.9
E	13.0	12.1	12.7	12.2	12.5	13.0	13.9	12.6	11.9	11.0	11.2	12.1	13.2	11.7	10.0
F	13.2	11.8	12.2	12.6	12.6	11.0	11.7	11.2	12.0	12.3	11.9	7.9	13.0	13.1	11.1

pH level

TABLE XIII. ZIRCONIUM-95 VALUES: CONTAINER versus DAY, Σ pH ACTIVITIES i.e. pCi/3 l

Container type	Day 7	Day 14	Day 28	Day 91	Day 210	Σx_{ij}^2	$\dfrac{(\Sigma x_{ij})^2}{5}$	$\bar{x}_{ij}$
A	32.3	36.3	37.4	32.7	35.8			34.9
B	36.7	35.8	35.9	33.3	40.4			36.4
C	35.3	35.1	36.4	38.4	39.0			36.8
D	35.2	35.0	36.2	37.3	41.8			37.1
E	37.8	37.7	38.4	34.3	34.9			36.6
F	37.2	36.2	34.9	32.1	37.2			35.5
Σx_{ji}^2						39526	39404	
$\dfrac{(\Sigma x_{ji})^2}{6}$						39425		
$\bar{x}_{ji}$	35.8	36.0	36.5	34.7	38.2			
pCi/1	11.9	12.0	12.2	11.6	12.7			

TABLE XIV. ZIRCONIUM-95 VALUES: DAY versus pH, Σ CONTAINER
ACTIVITIES, i.e. pCi/6 l

| Day | pH level | | | Σx_{ij}^2 | $\dfrac{(\Sigma x_{ij})^2}{3}$ | $\bar{x}_{ij}$ |
	1.8	3.0	N			
7	72.3	68.5	73.7			71.5
14	72.0	71.5	72.6			72.0
28	70.9	74.1	74.2			73.1
91	71.7	70.2	66.2			69.4
210	77.8	86.5	64.8			76.4
Σx_{ji}^2				79127	78850	
$\dfrac{(\Sigma x_{ji})^2}{5}$				78810		
$\bar{x}_{ji}$	72.9	74.2	70.3			
pCi/l	12.2	12.4	11.7			

only the interactions containers versus days and pH versus containers are
insignificant. The new residual variance therefore = $(3.521 + 2.882 + 6.758)/$
$(20 + 10 + 40) = 0.1880$ with 70 degrees of freedom. Using this new residual
variance, F ratios of 4.78 for the 'between containers' and 5.06 for the
interaction days versus pH are obtained and both are significant. Inspection
of the components of variance of the two remaining sources — 'between pH
levels' and 'between days' — shows that, with the σ_{CP}^2 and σ_{CD}^2 terms being
insignificant, the variance of the days versus pH interaction contains the
necessary components $(6\sigma_{PD}^2 + \sigma_0^2)$ to enable the significance of the $\sigma_P^2 + \sigma_D^2$
component to be assessed; using the variance of this interaction (0.9519),
F ratios of 18.6 and 50.1 for 'between pH' and 'between days' are obtained,
which are highly significant.

Figure 1 shows the variation of the two most significant variables —
pH levels and days of storage — using the means of the six container values,
together with the Day 0 level. The data indicate that activity values at all
pH levels decrease at the same rate until Day 28, when the rate of decrease
of the activity at the normal pH of seawater becomes more rapid.

There is a smaller, but still significant, difference in the variable
'between containers'; the mean activity levels for each container (irrespective
of pH and days of storage) are shown in Table XI, together with the surface
area/volume (SA/V) in cm^2/l. The 95% confidence limits for the differences
between two means = $\pm 2.01\sqrt{\dfrac{2 \times 0.1880}{15}} = \pm 0.32$. Container A is therefore
significantly better than containers C-F, and containers B-D are significantly
better than F.

The containers were of different materials and unfortunately the sizes
available were such that A and B were high-density polyethylene, C low-
density polythene, D polypropylene and E and F were PVC; it could therefore be argued that container materials rather than SA/V was the significant

TABLE XV. ZIRCONIUM-95 VALUES: pH versus CONTAINER, Σ DAY ACTIVITY, i.e. pCi/5 l

pH level	Container type						Σx_{ij}^2	$\dfrac{(\Sigma x_{ij})^2}{6}$	$\bar{x}_{ij}$
	A	B	C	D	E	F			
1.8	56.2	60.0	61.0	61.4	63.3	62.8			60.8
3.0	60.3	63.9	61.8	64.1	60.1	60.6			61.8
$\bar{x}$	58.0	58.2	61.4	60.0	59.7	54.2			58.6
Σx_{ji}^2							65 755	65 675	
$\dfrac{(\Sigma x_{ji})^2}{3}$							65 672		
$\bar{x}_{ji}$	58.2	60.7	61.4	61.8	61.0	59.2			
pCi/l	11.6	12.1	12.3	12.4	12.2	11.8			

DUTTON et al.

TABLE XVI. ZIRCONIUM-95 VALUES (INCLUDING DAY 210): ANALYSIS OF VARIANCE

Source	s.s.	d.f.	Mean square	Components of variance	F ratio
Between containers	6.01	5	1.20	$15\sigma_C^2 + 5\sigma_{CP}^2 + 3\sigma_{CD}^2 + \sigma_0^2$	
pH levels	6.52	2	3.26	$30\sigma_P^2 + 5\sigma_{CP}^2 + 6\sigma_{PD}^2 + \sigma_0^2$	2.14
days	13.19	4	3.30	$18\sigma_D^2 + 3\sigma_{CD}^2 + 6\sigma_{PD}^2 + \sigma_0^2$	2.17
Interactions					
Days versus pH	39.63	8	4.95	$6\sigma_{PD}^2 + \sigma_0^2$	4.29
Containers versus days	27.77	20	1.39	$3\sigma_{CD}^2 + \sigma_0^2$	
pH versus containers	9.92	10	0.992	$5\sigma_{CP}^2 + \sigma_0^2$	
Residual	43.06	40	1.08	σ_0^2	
Total	146.1	89			

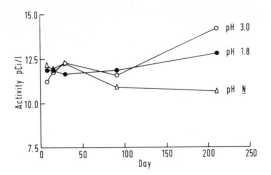

FIG. 3.   Effect of pH on the storage of ^{95}Zr.

factor. However, the variation in mean activity with the relative surface areas of the containers appeared significant and Fig.2 shows this relation-ship between the two variables; the Day 0 value of 9.6 pCi/l has been ascribed to SA/V = 0 (i.e. the expected mean activity for a container with no adsorption due to surface effect), and it can be seen that a hyperbolic-type function is produced consistent with the limits SA/V $\rightarrow \infty$, activity $\rightarrow$ 0. No good fit of functions was obtained when mathematical treatment was attempted.

Zirconium-95

The data obtained for ^{95}Zr are shown in Table XII; the groups of data used for the analysis of variance are shown in Tables XIII, XIV and XV, with the final calculations of variance in Table XVI. There is no significant difference between the containers, nor is there any significant interaction between containers versus days and between pH versus containers. The new residual variance = (27.77 + 9.92 + 43.06)/70 = 1.154. The days versus pH interaction appears significant, but this can confidently be ascribed to analytical imprecision at 210 days, and Fig.3 show how, at Day 210, the activity apparently increases for pH levels 1.8 and 3.0 and decreases for pH N. If the assumption is then made that the interaction is therefore insignificant, this term can be used to estimate the F ratio for the 'between pH levels' and 'between days' variables because the component of variance ($6 \sigma_{PD}^2 + \sigma_0^2$) is common to them all. F ratios thus calculated for these variables are 2.14 and 2.17 respectively. The variable 'between pH levels', therefore, is insignificant (F = 2.14 and P > 0.1); for the variable 'between days' an F ratio of 2.17 indicates 0.05 < P < 0.1, so that although this is highly insignificant, no positive effect can be ascribed to this variable.

The uncertainties described above are undoubtedly due to the analytical imprecision caused by radioactive decay over 210 days (^{95}Zr T$_{\frac{1}{2}}$: 65 days) and for this reason alone storage for such long periods should not be contemplated. Accordingly, the analysis of variance calculations were carried out for Days 7, 14, 28 and 91, and the data are summarized in Table XVII. None of the effects are significant.

TABLE XVII. ZIRCONIUM-95 VALUES (EXCLUDING DAY 210):
ANALYSIS OF VARIANCE

Source	s.s.	d.f.	Mean square	F ratio
Between containers	4.96	5	0.992	
pH levels	0.18	2	0.090	
days	3.66	3	1.22	1.24
Interactions				
Days versus pH	6.20	6	1.03	
Containers versus days	8.82	10	0.882	
pH versus containers	16.76	15	1.12	
Residual	30.12	30	1.00	
Total	70.7	71		

Day 0 activity levels of 12.3, 11.7 and 12.0 pCi/l (mean 12.0 pCi/l)
were obtained; this is not significantly different from the mean of the
stored values (12.1 ± 1.0 pCi/l). Over the 210 days of storage, therefore,
the activity did not drop by $2\sqrt{2 \times 1.0^2}$ (95% confidence) = 2.8 pCi/l,
equivalent to ≯ an average of 0.11% per day.

CONCLUSIONS

For the particular environment investigated there were no significant
losses of [106]Ru and [95]Zr in filtered seawater from a variety of containers
and at pH levels between 1.8 and ≈ 8, the natural pH of seawater. Calculated
limits (95% probability) were ≯ 0.06% per day (average) and ≯ 0.11% per day
(average) respectively over 210 days. No suitable storage conditions have
been established for [144]Ce, and it is hoped to test the storage stability that
lower pH values in the near future.

ACKNOWLEDGEMENTS

We wish to thank all those who helped with the sample collection and
treatment, both during the 1969 cruise of RV CORELLA and at Whitehaven
during 1970, especially A.K. Steele and C.W. Baker. Our thanks due also
to our other colleagues at the Fisheries Laboratories, Lowestoft, for
technical discussions and assistance in the preparation of the manuscript.
The work was supported in part by an IAEA contract, and we wish to
thank Dr. Fukai of the Monaco Laboratory for his continual encouragement
and support. The valuable suggestions of Dr. V.T. Bowen of WHOI are
also gratefully acknowledged, as is his help in obtaining the US containers.

APPENDIX I

THE DETERMINATION OF RUTHENIUM-106 IN SEAWATER

INTRODUCTION

The method closely follows the standard procedure recommended for natural waters [8] and involves carrier addition, isotopic equilibration by alkaline oxidation to the perruthenate and extraction of ruthenium tetroxide into carbon tetrachloride. An alkaline backwash is followed by reduction with methanol to ruthenium dioxide, which is dissolved in hydrochloric acid and further reduced to ruthenium metal with magnesium powder. The metal is weighed to determine the chemical yield and β-counted through an absorber (to eliminate any ^{103}Ru interference) to determine the ^{106}Ru content.

REAGENTS

1.   Potassium periodate, potassium persulphate, potassium hydroxide, nitric acid (SG 1.42), hydrochloric acid (SG 1.18), carbon tetrachloride and methanol — analytical reagent grades.
2.   Sodium hydroxide solution — 12 wt/vol (%) analytical reagent grade in distilled water.
3.   Magnesium powder — laboratory reagent grade.
4.   Ruthenium carrier solution. Ruthenium chloride is dissolved in concentrated hydrochloric acid to make a solution approximately 10 mg Ru/ml. The solution is standardized by reducing a 5 or 10 ml aliquot with magnesium powder and weighing the ruthenium metal produced; the method is the same as that detailed in the experimental procedure.
5.   Standardized ^{106}Ru solution, obtainable from the Radiochemical Centre, Amersham. The solution is accurately diluted, if necessary, until the ^{106}Ru level is in the region of 10^3 pCi/g.

SPECIAL APPARATUS

Gas-flow thin (80 μg/cm^2) end-window (5 cm diameter) Geiger counter: the Tracerlab Omniguard was used.

PROCEDURE

1.   Shake the container and transfer a 4 litre portion to a 5 litre beaker.
2.   Add 2 ml ruthenium carrier.
3.   Add 40 g potassium persulphate, 80 g potassium hydroxide pellets, and 20 g potassium periodate.
4.   Warm to 90°C; maintain at that temperature for 15 minutes and allow to cool.

5.   Add 100 ml carbon tetrachloride, with continuous stirring, and 200 ml concentrated nitric acid. This addition results in a pH of about 1 and ensures complete dissolution of any precipitate but does not affect the ruthenium extraction, which is normally carried out at pH 4-6. Stir for 20-30 minutes and allow to settle.

6.   Remove and retain the carbon tetrachloride layer, add a further 50 ml carbon tetrachloride to the aqueous phase and stir again. Remove as much as possible of the carbon tetrachloride, bulk the organic extracts and discard the aqueous phase.

7.   Shake the bulked carbon tetrachloride extracts with 200 ml distilled water, allow to settle and discard the aqueous phase. Backwash the carbon tetrachloride with 90 ml 12% NaOH solution and transfer the aqueous phase to a 100 ml centrifuge tube; discard the carbon tetrachloride.

8.   Add 0.5 ml methanol to the alkaline backwash, which is then stood in a boiling water bath until the ruthenium dioxide coagulates. Centrifuge and discard the supernate.

9.   Dissolve the precipitate in a few drops of concentrated hydrochloric acid — gentle heating over a flame is sufficient.

10. Add about 0.6 g magnesium powder, swirling gently; when the reduction is complete, dissolve all excess magnesium by the careful addition of small quantities of concentrated hydrochloric acid.

11. Dilute to 50 ml with distilled water and filter on a tared paper (Whatman 42, 24 mm diameter). Wash well with water, dry in an oven at 105°C, and weigh to determine the chemical yield.

12. Mount the paper on a steel planchette and measure the count rate of the sample through a 50 mg/cm^2 absorber.

13. Measure the background by counting a ruthenium blank. This is prepared by reducing a 2 ml aliquot of the ruthenium carrier solution with magnesium powder and mounting on a filter paper and steel planchette by the procedure described above.

14. Prepare standard sources of [106]Ru, produced in a similar manner to the above procedure, using a suitable aliquot of the standardized [106]Ru solution (containing 200-300 pCi [106]Ru) with 4 litres of inactive seawater.

15. Convert the count rate of the samples (after correction for background and chemical yield) to pCi/l seawater by comparison with the count rate of the standards produced in item 14, and correct the activity thus obtained to that present on the reference date (Day 0).

THE DETERMINATION OF ZIRCONIUM-95 AND
CERIUM-144 IN SEAWATER

INTRODUCTION

Zirconium-95 and cerium-144 are coprecipitated on Ce^{III}, lanthanum and Fe^{III} hydroxides. After dissolution in perchloric acid and evaporation to dryness to remove radioruthenium, the residue is dissolved in concentrated hydrochloric acid, and ^{144}Ce and ^{95}Zr are separated using anion exchange. Cerium-144 is not adsorbed on the column and is precipitated as ceric periodate (ensuring a separation from yttrium and the rare earths). Zirconium-95 is eluated from the column with a hydrochloric/hydrofluoric acid mixture and is coprecipitated, with a further addition of lanthanum carrier, as lanthanum fluoride. Cerium recovery is measured by stable cerium addition; the yield of zirconium is estimated by measuring the recovery of added stable iron.

REAGENTS

1.   Hydrochloric acid (SG 1.18), nitric acid (SG 1.42), perchloric acid (SG 1.54), hydrofluoric acid (48 wt/vol (%)), ammonia solution (SG 0.88) — analytical reagent grades.
2.   Saturated ammonium fluoride solution, $2\underline{M}$ sodium bromate solution — analytical reagent grades dissolved in distilled water.
3.   Saturated potassium periodate (analytical reagent grade) solution in 20% v/v nitric acid.
4. $0.02\underline{M}$ HF/$9\underline{M}$ HCl solution: 1 g hydrofluoric acid and 750 ml hydrochloric acid made up to 1 litre with distilled water.
5.   Anion exchange resin: Biorad AG 1 × 10 (Cl⁻), 100-200 mesh equilibrated with hydrochloric acid.
6.   Mixed carrier solution: 2 mg/ml each of La^{III}, Ce^{III} and Fe^{III} as chlorides in $0.5\underline{M}$ HCl. The cerium and iron contents must be accurately known because they are used for yield measurement.
7.   Lanthanum carrier: 10 mg/ml La^{III} as the chloride in $0.5\underline{M}$ HCl.
8.   Rare earth hold-back carrier: 2 mg/ml each of Nd^{III}, Sm^{III} and Y^{III} as chlorides in dilute acid.
9.   Ruthenium carrier solution — see Appendix I.

SPECIAL APPARATUS

Gas-flow thin (80 $\mu g/cm^2$) end-window (5 cm diameter) Geiger counter: the Tracerlab Omniguard was used.

PROCEDURE

1.   Shake the container and transfer a 4 litre portion to a 5 litre beaker.
2.   Add 40 ml nitric acid, 5 ml of the mixed carrier solution, and mix well.
3.   Add 100 ml ammonia solution and allow to stand overnight.
4.   Siphon off the supernate, until approximately 300 ml is left, and transfer
to a 600 ml centrifuge bottle with water washing.
5.   Centrifuge and reject the supernate. Dissolve the precipitate in 10 ml
hydrochloric acid, dilute to 300 ml with water and reprecipitate with 25 ml
ammonia solution.
6.   Centrifuge and reject the supernate. Dissolve the precipitate in 50 ml
perchloric acid and transfer into a 500 ml beaker with a water wash. Add
1 ml ruthenium carrier, evaporate to dryness and bake until fuming has
ceased.
7.   Dissolve the residue in the minimum quantity of $12\underline{M}$ hydrochloric acid.
8.   Prepare an anion exchange column 90×6 mm diameter of Ag 1×10
resin, 100-200 mesh, equilibrated with $12\underline{M}$ HCl and load the dissolved
residue on the column, transferring with washes of $12\underline{M}$ hydrochloric acid.
9.   Wash the column with 25 ml $12\underline{M}$ hydrochloric acid, collect the effluent
in a 500 ml beaker and reserve it for cerium analysis (Sections 17-25).

Zirconium analysis

10.   Elute the zirconium from the column with 12 ml of $0.02\underline{M}$ HF/$9\underline{M}$ HCl
solution and collect in a 25 ml plastic centrifuge tube.
11.   Add 1 ml of the lanthanum carrier solution (10 mg La^{III}) and 2 ml
saturated ammonium fluoride solution.
12.   Stir well and centrifuge for 5 minutes at 500 rev/min.
13.   Filter the precipitate on a paper, taking care to effect a quantitative
transfer.
14.   Wash the column with 25 ml $0.5\underline{M}$ HCl and measure the iron content
of this eluate by the standard atomic absorption spectrophotometric
procedure. Compare this with the added amount of iron (item 2) to determine the chemical yield of iron (and hence zirconium).
15.   Measure the β-activity of the separated ^{95}Zr sources.
16.   Measure the β-activity of a blank tray prepared by treating only stable
carriers with the reagents.
17.   Prepare counting standards of ^{95}Zr by evaporating a known quantity
of ^{95}Zr solution almost to dryness and continuing as in items 7, 8, 10-13,
15 and 16. These standards must be prepared and counted at the same
time as the sample, because of ^{95}Nb grow-in.
18.   Convert the count rate of the sample (after correction for background
and chemical yield) to pCi/l seawater by comparison with the count rates
of the standards. Correct the activity to that present on the reference
date (Day 0).

Cerium analysis

19.   Add 1 ml of rare earth hold-back carrier to the effluent from item 9
and evaporate to dryness on a hotplate.

20. Dissolve the residue in 25 ml 20% v/v nitric acid and add 2 ml 2M sodium bromate solution and 25 ml saturated potassium periodate solution in 20% v/v nitric acid.

21. Heat to coagulate the precipitate and filter on to a tared filter paper, wash well with water and dry in an oven at 105°C, reweigh to obtain the weight of cerium periodate $[Ce(IO_4)_4]$ recovered, and mount on a stainless steel counting tray.

22. Measure the β-activity due to [144]Ce by counting the tray on the β^- counter through a 50 mg/cm^2 absorber (to eliminate any interference from [141]Ce).

23. Measure the β-activity of a blank tray.

24. Prepare counting standards of [144]Ce using a known amount of [144]Ce, with 5.0 ml of the cerium carrier solution, and carrying out the procedures in items 19 to 23 inclusive.

25. Convert the count rate of the sample (after correction for background and chemical yield) to pCi/l of seawater by comparison with the count rate of the standards, and correct the activity to that present on the reference date (Day 0).

REFERENCES

[1] KENNY, A.W., MITCHELL, N.T., "United Kingdom waste management policy", Management of Low- and Intermediate-level Radioactive Waste (Proc. Symp. Aix-en-Provence, 1970), IAEA, Vienna (1970) 69.

[2] PRESTON, A., DUTTON, J.W.R., "Review of methods for the determination of selected cobalt nuclides in marine materials", Reference Methods for Marine Radioactivity Studies, Technical Reports Series No. 118, IAEA, Vienna (1970) 223-41.

[3] ANDERSEN, N.R., HUME, D.N., Determination of barium and strontium in sea water, Anal. Chim. Acta 40 (1968) 207.

[4] ROBERTSON, D.E., The adsorption of trace elements in sea water on various container surfaces, Anal. Chim. Acta 42 (1968) 533.

[5] INTERNATIONAL ATOMIC ENERGY AGENCY, Methods of Surveying and Monitoring Marine Radioactivity, Safety Series No. 11, IAEA, Vienna (1965) 95.

[6] WORTHING, A.G., GEFFNER, J., Treatment of Experimental Data, Wiley, New York (1948) 342.

[7] YATES, F., The analysis of replicated experiments when field results are incomplete, Emp. J. Exp. Agric. 1 (1933) 129.

[8] WORLD HEALTH ORGANIZATION, Methods of Radiochemical Analysis, FAO, IAEA, and WHO, Geneva (1966) 163.

INTERCALIBRATION OF METHODS
FOR RADIONUCLIDE MEASUREMENTS
ON A SEAWEED SAMPLE

R. FUKAI, S. BALLESTRA
International Laboratory of Marine Radioactivity,
Oceanographic Museum,
Monaco-Ville,
Principality of Monaco

Abstract

INTERCALIBRATION OF METHODS FOR RADIONUCLIDE MEASUREMENTS ON A SEAWEED SAMPLE.
    The results obtained in the intercalibration exercise of radionuclide measurements on a seaweed sample
are surveyed. Forty-seven laboratories of world-wide distribution participated in this exercise during 1971-72.
It was demonstrated by the survey that the comparability of the reported results on strontium-90 and caesium-137,
obtained by using various methods, is fairly satisfactory, except for a few widely deviated results. On the
other hand, the variation of the reported results on potassium-40, ruthenium-106 and cerium-144 are
considerably larger, even though the majority of results had been obtained by using non-destructive γ-spectrometry with Ge(Li) detectors. This indicates that emphasis should be placed on a proper control of calibration
procedures of counting instruments for these radionuclides. The probable concentrations estimated on the
basis of statistical treatments of the reported results, are presented for the major radionuclides, potassium-40,
strontium-90, ruthenium-106, caesium-134, caesium-137 and cerium-144, as well as low-level radionuclides,
cobalt-60, zinc-65, zirconium-niobium-95 and silver-110m. The preparation and the homogeneity of the
distributed samples are also described in detail.

1. INTRODUCTION

    Prospective views on accelerating developments in the field of nuclear
industry in the near future have induced increasing public concern on the
safe control of radioactive waste releases to the environment and their
eventual impact on human health and welfare. The environmental monitoring
of radioactive substances plays an essential role in ensuring safe operation
of nuclear installations for both the public and the competent authorities.
Since the sea represents a sector of the environment in which all nations
are interested, internationally comparable results of the monitoring of the
marine environment are especially desirable. As the techniques of
monitoring are still rapidly developing and various methods of measurements
are consequently employed by different laboratories, the necessary assurance
of the comparability of the results can only be provided either by group
intercalibration analyses of aliquots of homogeneous samples or by analyses
of standard reference material. The experience gained by the intercalibration
exercise of fission product measurements on seawater samples [1] has
indicated further need for such an exercise on different sorts of matrices.
Since the normal practice for the monitoring of the marine environment
involves measurements of radionuclides on biological samples rather than
seawater samples, a seaweed sample was chosen as the second intercalibration sample during 1971-72. In the present paper a survey is made of
the results reported by participating laboratories in this intercalibration
exercise.

It should be emphasized that the material presented in this paper has
been produced by the joint efforts of all participating laboratories with the
common aim of improving future comparability of the results within the
group and to contribute to the whole scientific community by doing so.

2.   COLLECTION AND PREPARATION OF THE SEAWEED SAMPLE

Since radiochemical separations have to be employed for some radionuclides and by certain laboratories, the preparation of samples spiked
with known amounts of radionuclides was avoided in order to ensure that
chemical forms of radionuclides present in the dispatch sample were
identical to those in samples from environmental origin. Therefore, the
effort was directed to the field collection of sufficient amounts of seaweed
that was expected to be contaminated with radionuclides at monitoring levels.
The field collection was carried out by the British Fisheries Radiobiological
Laboratory, Lowestoft, with the financial support of the IAEA. About 50 kg
of seaweed (Fucus serratus) was collected from a coastal zone during early
1971, stirred in tap water in a large plastic tank to remove excess silt, etc.
and then transferred in batches to a domestic washing machine where it was
again washed and spin-dried. The well-washed seaweed was then weighed
and dried in an oven at 100 - 105°C; the percentage of the dried matter
relative to the wet weight was 17.8%. The dried seaweed was placed in a
polyethylene bag and broken into small flakes using a rubber mallet; these
flakes were then coarsely ground by means of a mill, mixed again in a
polyethylene bag and sub-sampled for preliminary homogeneity tests at the
Lowestoft Laboratory. The remaining dried seaweed of about 10 kg was
dispatched to the Monaco Laboratory. At Monaco the dried seaweed was
further ground in a porcelain ball-bearing mill to finer sizes, mechanically
homogenized and twice sieved with nylon plankton nets having 20-mesh and
200-mesh sizes, respectively. The fraction between 20 and 200-mesh was
taken and divided into polyethylene bottles, each of which contained
50 ± 0.1 g of dried and ground seaweed. At this stage the sample contained
4.6% moisture, which can be removed by drying at 105°C. The seaweed
sample was code numbered as AG-I-1.

3.   HOMOGENEITY OF THE SAMPLE

Preliminary homogeneity tests carried out at the Lowestoft Laboratory
by semi-quantitative γ-spectrometry on major peak areas showed that the
levels of major γ-emitting radionuclides in 7 sub-samples taken (70 ± 3 g
each) varied less than $\pm 5\%$ (1σ - standard deviation) from the mean.
After further grinding, a second homogenization and sieving, the
homogeneity of the sample was tested once again at the Monaco Laboratory.
Before starting, it was confirmed that no loss of radionuclides occurred by
drying the seaweeds at 105 - 110°C. Five grams of the dried seaweed were
put in a plastic counting tube and counted repeatedly 5 times for 100 min,
using a 7.5×7.5 cm NaI well crystal coupled to a 200-channel analyser.
The results of counting of major peak areas are shown in Table I. The
standard deviation σ_c for each peak area represents the counting statistics
including those due to variations of background, electric current, etc.

TABLE I. RESULTS OF HOMOGENEITY TESTS ON AG-I-1 BY COMPARING REPRESENTATIVE PEAK AREAS OF γ-SPECTRA

Channel range	12 - 16	55 - 65	66 - 85	86 - 90	121 - 140	161 - 175
Representative peak isotope and energy	^{144}Ce 0.134 MeV	^{106}Ru-Rh 0.51 MeV	^{137}Cs 0.66 MeV	^{95}Zr-Nb 0.75 MeV	^{106}Ru-Rh 1.05, 1.13 MeV	^{40}K 1.46 MeV
Repeated counting on 1 sub-sample (counts/100 min/g)	1442	1938	5105	581	862	404
	1482	1970	5051	613	853	396
	1473	1981	5087	594	863	390
	1467	1965	5059	634	847	379
	1446	1955	5036	585	851	387
Mean	1462	1962	5068	601	855	391
σ_c (single det.)	± 1.2%	± 0.8%	± 0.6%	± 3.7%	± 0.8%	± 2.4%
Counting on 5 different sub-samples (counts/100 min/g)	1434	1954	5091	607	856	369
	1473	1981	5087	594	863	390
	1466	1972	5008	579	849	368
	1481	2005	5041	596	870	389
	1491	1990	5069	583	851	393
Mean	1469	1980	5059	592	858	382
σ_s (single det.)	± 1.5%	± 1.0%	± 0.7%	± 1.9%	± 1.0%	± 3.2%
Residual standard deviation $\sigma_r = \sqrt{\sigma_s^2 - \sigma_c^2}$	± 0.9%	± 0.6%	± 0.4%	-	± 0.6%	± 2.1%

during the period of counting. In Table I the results are also given for the counting of 5 different sub-samples under similar counting conditions. The standard deviation σ_s is considered to represent the variation due to sample composition plus overall counting statistics. Accordingly, the residual standard deviation σ_r, estimated as shown in Table I, can be regarded as variations due to sub-sample inhomogeneity, assuming constant background throughout the counting period. As can be seen from the table, the values for σ_r do not exceed ± 1%, except for the peak areas for zirconium-niobium-95 (in this case $\sigma_c > \sigma_s$) and potassium-40, both of which have relatively low count rates. From these values it is reasonable to conclude that levels of major γ-emitters at a sample size of 5 g vary less than ± 1% (1 σ - standard deviation) due to a difference in the composition and that overall variation should not exceed ± 2% standard deviation for the major artificial radionuclides.

In Table II the results are given of repeated determinations carried out by the Monaco Laboratory and the Woods Hole Oceanographic Institution of strontium-90, ruthenium-106, caesium-134, caesium-137 and cerium-144. The analytical methods adopted at Monaco were those described by Bowen [2] for strontium-90, by FAO/IAEA/WHO [3] for ruthenium-106 with slight modification in counting steps, by Fukai et al. [4] for caesium-134 and 137 and by γ-spectrometry on hydroxide precipitate for cerium-144. At Woods Hole a similar method to that of Monaco was used for strontium-90,

TABLE II. RESULTS OF HOMOGENEITY TESTS ON AG-I-1 BY
DETERMINATIONS OF INDIVIDUAL RADIONUCLIDES

Laboratory	Sample bottle No.	Quantities found[a] (pCi/g-dried matter)[b]				
		^{90}Sr	^{106}Ru	^{134}Cs	^{137}Cs	^{144}Ce
MONACO[c]	83	9.9	-	-	-	-
	84	9.2	-	-	-	-
	86	10.1	138	11.1	70.5	19.5
	86	10.0	-	5.9	74.8	18.0
	86	9.9	-	6.8	72.9	-
	87	11.4	143	9.3	74.3	20.0
	87	10.2	150	7.4	71.3	17.9
	87	10.7	-	11.5	70.3	-
	91	10.3	136	-	-	-
	91	10.0	146	-	-	-
	Mean	10.2	143	8.7	72.4	18.9
	σ	±0.6	±6	±2.3	±1.8	±1.0
	(Single det.)	(5.9%)	(4.2%)	(26%)	(2.5%)	(5.0%)
	Chauvenet's range	±1.2	±10	±4.0	±3.1	±1.5
WHOI[d]	38	9.8 ⎫	145	10.3	77.5	16.2
	38	9.5 ⎭				
	39	10.2 ⎫	144	10.7	78.9	17.9
	39	10.6 ⎭				
	59	10.0 ⎫	146	10.4	77.6	16.8
	59	10.7 ⎭				
	Mean	10.1	145	10.5	78.0	17.0
	σ	±0.5	±1	±0.2	±0.8	±0.9
	(Single det.)	(5.0%)	(0.7%)	(1.9%)	(1.0%)	(5.3%)

[a] Activities normalized on 1 Jan. 1972.
[b] Dried at 105 - 110°C.
[c] International Laboratory of Marine Radioactivity, IAEA, Monaco.
[d] Woods Hole Oceanographic Institution.

while other radionuclides were measured by γ-spectrometry using Ge(Li) detectors on the dried seaweed. Due to radiochemical procedures involved in the determinations the variations of the results of these analyses are larger than those obtained by the semi-quantitative γ-measurements. These results indicate, however, that there is no reason to believe that greater inhomogeneity exists with respect to the individual radionuclides analysed than was shown by the semi-quantitative γ-measurements.

4. PROCEDURE FOR INTERCALIBRATION

The bottles, containing 50 g of dried and ground seaweed, were distributed to 55 institutions (including the IAEA's Seibersdorf Laboratory) from 25 Member States of the IAEA. The participants were informed about the major radionuclides present in the sample and the approximate homogeneity; they were instructed to dry the sample at approximately 105°C prior to the start of the analysis. At the stage of result reporting the participants received standard forms, which included a report on the analytical data as well as information on instrumentation, calibration, calculation and analytical procedures. The reference date for the reporting radioactivity was set as 1 January 1972.

Finally, 45 institutions from 21 Member States and 2 laboratories of the IAEA reported their results. A list of the institutions that participated in this intercalibration exercise is given as an Annex with names of chief investigators and their collaborators.

5. SURVEY OF THE RESULTS

Data treatment

Table III is a compilation of the reported results of 10 frequently reported radionuclides (potassium-40, cobalt-60, zinc-65, strontium-90, zirconium-niobium-95, ruthenium-106, silver-110m, caesium-134, caesium-137 and cerium-144). All values given in this table were taken from the original reports of the participants with as little change as possible; when the reported activity had not been normalized to the reference date of 1 January 1972 the decay was corrected; moisture correction was applied when the sample had not been oven-dried before the start of the analysis; in cases where 2 or more results had been reported for one radionuclide, an arithmetic average was calculated and presented in the table. The average values are underlined in the table. The errors presented in Table III represent in most cases standard deviations in counting statistics of the presented values; for the average values calculated for 2 or 3 results of one radionuclide the errors were computed from individual errors given, while standard deviations of the average values were computed from the individual results in cases where 4 or more results were available.

Table IV gives the results of less frequently reported radionuclides. Except for strontium-89, these radionuclides were measured by γ-spectrometry using Ge(Li) detectors. Considering the relatively short half-lives of some of these radionuclides and the length of the storage time between receipt and dispatch of the sample, the values for strontium-89,

TABLE III. RESULTS OF MEASUREMENTS ON TEN FREQUENTLY REPORTED RADIONUCLIDES IN THE SEAWEED SAMPLE AG-I-1 BY DIFFERENT INSTITUTIONS
(pCi/g-dried matter at 105°C)

Code No. of laboratory	^{40}K	^{60}Co	^{65}Zn	^{90}Sr	^{95}Zr-Nb
1	56.7 ± 19.9	4.95 ± 2.31	-	-	2.85 ± 2.40
2	25.1 ± 5.0	0.4	0.1	-	N.D.
3	35	2.5	2.0	-	4.2
4	55	2.0 ± 0.6	-	9.5	3.1 ± 0.7
5	-	-	-.	-	-
6	-	-	-	9.85 ± 0.49	-
7	36.5 ± 0.5	1.88 ± 0.04	1.3 ± 0.3	-	1.98 ± 0.12
8		2.03 ± 0.16	1.67 ± 0.47	9.09 ± 0.15	2.23
9	-	2.1 ± 0.4	-	10.2 ± 1.5	-
10	-	-	-	13.4 ± 0.4	-
11	67 ± 2	-	-	7.0 ± 1.0	-
12	-	-	-	8.4 ± 0.5	-
13	-	1.83 ± 0.09	1.0 ± 0.3	10.02 ± 0.25	1.91 ± 0.24
14	35.4 ± 4.2	2.8 ± 0.2	-	11.05 ± 0.13	1.1 ± 0.3
15	-	-	-	10.6 ± 0.4	N.D.
16	50.6 ± 3.5	2.15 ± 0.32	2.47 ± 0.80	9.74 ± 0.48	4.00 ± 2.03
17	-	-	-	-	-
18	-	2.14 ± 0.06	-	-	-
19	30.7 ± 1.2	-	-	-	-
20	37.1 ± 3.8	2.2 ± 0.2	1.1 ± 0.3	9.6 ± 0.3	3.0 ± 0.4
21	33 ± 1	1.8 ± 0.1	-	-	2.21 ± 0.23
22	41 ± 4	-	-	9.3 ± 0.1	1.94 ± 0.51
23	-	-	-	-	0.347 ± 0.004
24	79.2 ± 2.5	4.5 ± 0.4	-	9.9	8.9 ± 1.0
25	11.3 ± 1.0	10.8 ± 1.1	3.7 ± 1.7	-	N.D.
26	27 ± 2	-	-	-	-
27	-	-	-	-	5.45
28	-	-	-	-	-
29	-	-	-	5.9 ± 0.5	9 ± 5
30	-	-	-	10.10 ± 0.40	-
31	-	1.60 ± 0.10	-	10.34 ± 0.16	2.4 ± 0.2
32	35 ± 3	1.6 ± 0.2	-	10.5 ± 1.0	<4.3
33	-	-	-	-	-
34	-	-	-	-	-
35	37 ± 2	-	-	10.2 ± 0.2	3.6 ± 0.8
36	-	-	-	-	-
37	38 ± 2	1.8 ± 0.2	-	-	-
38	-	-	-	15.6 ± 1.1	-
39	-	-	-	7.77 ± 0.09	-
40	-	-	-	-	50 ± 14
41	-	-	-	-	-
42	58 ± 6	-	-	-	-
43	-	-	-	-	-
44	34.3 ± 0.8	1.74 ± 0.09	-	-	-
45	39 ± 2.1	11 ± 0.9	-	2.4 ± 11.9	-
46	66 ± 7	-	25 ± 4	11.3 ± 0.2	6 ± 2
47	-	-	-	9.5 ± 0.9	-

Values underlined: Average values of 2 or more results.
N.D.: Not detected or not determined due to too low concentration.

TABLE III (cont.)

Code No. of laboratory	^{106}Ru	^{110}Ag^m	^{134}Cs	^{137}Cs	^{144}Cs
1	157 ± 14	-	N.D.	75.0 ± 5.1	7.9 ± 4.7
2	162.4 ± 32.5	-	14.9 ± 3.0	69.7 ± 13.9	36.9 ± 7.4
3	138	13.1	9.6	77	11.0
4	120 ± 2	1.6 ± 0.6	10 ± 2	60 ± 1	16 ± 1
5	846	-	715	5047	206
6	-	-	-	33.63 ± 1.68[a]	-
7	153 ± 9	1.67 ± 0.01	12.4 ± 1.2	75.2 ± 1.7	19.0 ± 0.6
8	158 ± 2	1.73 ± 0.28	10.9 ± 0.2	81.2 ± 0.6	19.8 ± 0.7
9	148 ± 6	1.3 ± 0.7	12 ± 1	72 ± 3	22 ± 3
10	189 ± 23	-	10.25 ± 0.71	78.6 ± 2.3	24.5 ± 3.6
11	138 ± 1	-	11.9 ± 0.2	70 ± 0.7	8.5 ± 0.3
12	107.8 ± 4.3	-	-	68.8 ± 4.0[a]	14.9 ± 0.9
13	145 ± 2	-	10.47 ± 0.23	78.2 ± 0.3	17.1 ± 0.4
14	99.9 ± 1.8	-	12.9 ± 5.1	74.9 ± 0.5	14.4 ± 1.0
15	159 ± 6	-	9 ± 1	70 ± 1.5	20 ± 2
16	188.1 ± 11.3	1.77 ± 0.53	13.4 ± 0.8	96.1 ± 4.8	26.9 ± 8.7
17	153 ± 3.7	-	-	-	-
18	-	1.63 ± 0.01	11.6 ± 0.2	73.3 ± 0.4	-
19	117.5 ± 1.8	-	8.7 ± 0.3	71.7 ± 0.3	17.0 ± 0.9
20	151 ± 2	1.5 ± 0.3	10.2 ± 0.2	79.0 ± 0.5	18.5 ± 0.9
21	126 ± 4	1.21 ± 0.12	8.5 ± 0.1	65.7 ± 0.2	14.7 ± 2.8
22	120.5 ± 7.1	-	9.3 ± 0.4	68.2 ± 0.9	14.2 ± 0.9
23	-	-	-	79.25 ± 6.27	-
24	72.0 ± 1.0	-	-	69.0 ± 0.6	8.0 ± 1.0
25	123 ± 7.0	-	9.7 ± 0.7	81.8 ± 2.2	20.4 ± 2.8
26	28 ± 2	-	10 ± 1	72 ± 3	-
27	-	-	-	-	-
28	103 ± 7	-	-	80 ± 8	-
29	74 ± 6	-	9.2 ± 1	71 ± 4	18 ± 3
30	-	-	-	89.0 ± 5.0[a]	-
31	145.7 ± 1.2	-	10.35 ± 0.15	77.6 ± 0.3	16.8 ± 0.7
32	130 ± 13	-	8.7 ± 0.7	64 ± 6	-
33	339 ± 3	-	-	-	-
34	155.7 ± 4.5	-	9.1 ± 0.8	84.1 ± 3.6	25.0 ± 3.4
35	143 ± 3	2.5 ± 0.5	8.7 ± 0.9	72.4 ± 0.8	18.9 ± 0.5
36	127.7 ± 1.4	-	11.7 ± 0.1	82.4 ± 0.4	18.2 ± 0.4
37	133 ± 7	1.1 ± 0.2	10 ± 1	75 ± 4	16 ± 1
38	-	-	-	-	-
39	-	-	-	74.6 ± 0.5[a]	-
40	-	-	-	-	-
41	167.32 ± 2.41	-	-	72.7 ± 0.77	-
42	-	-	-	84 ± 8	-
43	-	-	-	31.6 ± 0.4[a]	-
44	126 ± 2	1.70 ± 0.15	10.2 ± 0.2	76.4 ± 2.3	15.9 ± 0.5
45	4800 ± 450	-	8 ± 0.3	73 ± 0.7	175 ± 22
46	79 ± 20	-	6 ± 3	88 ± 9	-
47	-	-	-	-	-

Values underlined are the average values of 2 or more results.

N.D.: Not detected or not determined due to too low concentration.

[a] Values represent ^{134}Cs + ^{137}Cs.

TABLE IV.   RESULTS OF MEASUREMENTS ON LESS FREQUENTLY
REPORTED RADIONUCLIDES IN THE SEAWEED SAMPLE AG-I-1

Radionuclide	Half-life	Concentration reported (pCi/g-dried matter)	Code No. of laboratory
^{22}Na	2.58 years	0.503 ± 0.117 0.56 ± 0.20 0.3 ± 0.1	8 16 20
^{54}Mn	314 days	0.046 ± 0.013 7 ± 2	7 46
^{57}Co	267 days	0.4 ± 0.1	4
^{89}Sr	50.4 days	0.0 ± 1.5 <u>10.5 ± 2.6</u> 4	10 11 20
^{103}Ru	40 days	1.4 ± 0.8 N.D.	14
^{125}Sb	2.7 years	6.8 ± 3.1 0.67 ± 0.18 0.82 ± 0.11 4 ± 1.4	4 7 44 45
^{131}I	8.05 days	42.920 ± 5.65	25
^{141}Ce	32.5 days	0.9 ± 0.5	14
^{154}Eu	16 years	0.65 ± 0.06 0.11 ± 0.01 0.8 ± 0.1 1.11 ± 0.04 6 ± 1.6	7 21 37 44 45
^{155}Eu	1.7 years	1.2 ± 0.2 1.5 ± 0.2 1.5 ± 0.1 1.6 ± 0.2 8 ± 2.5	4 7 21 37 45
^{214}Bi	19.7 min	3.2 ± 1.5	24
^{226}Ra	1620 years	9.1 1.4 ± 0.3	4 14
^{228}Th	1.9 years	2.5 ± 0.2	45
^{232}Th	1.4×10^{10} years	<u>5.8 ± 0.4</u>	24

Values underlined are the average value of 2 or more results.
N.D.: Not detected.

ruthenium-103, iodine-131 and cerium-141 seem to be questionable. In addition to artificially produced radionuclides, Table IV also includes naturally occurring radionuclides from the uranium and thorium decay series.

Some of the laboratories measured transuranic elements such as plutonium-238 and 239 or americium-241. These results are incorporated in the report on the intercalibration of measurements of transuranic elements [5].

Scatter of the data

The overall averages of the ten frequently reported radionuclides are given in Table V with maximum and minimum values reported. To obtain overall averages all available values were arithmetically averaged, regardless of their wide variations. Consequently, the average values given in Table V have little statistical significance because of the wide ranges of variation, although the standard deviations of these values give a good idea of the degree of scatter for the individual radionuclides given. Of the 10 radionuclides presented, the standard deviation of 7 radionuclides (zinc-65, zirconium-niobium-95, ruthenium-106, silver-110m, caesium-134, caesium-137, and cerium-144) exceeds ±20% and that of 5 radionuclides (zinc-65, zirconium-niobium-95, ruthenium-106, caesium-134 and caesium-137) exceeds ±40%. The maximum and minimum values are different by a factor of nearly 200 for zirconium-niobium-95, ruthenium-106, and caesium-137. These large variations for caesium-134, caesium-137 and cerium-144 are caused, however, by exceptionally high values reported by one laboratory for these radionuclides. By removing these values from the averaging procedures one obtains standard deviations of less than ±10% for these radionuclides.

The total number of reported results for each radionuclide is not sufficiently large to allow significant statistical deductions on the distribution of the reported values over certain concentration ranges. Nevertheless, an attempt is made to illustrate the frequency of appearance of these values over concentration ranges by histogram expression. In Figs 1–3 per cent frequency histograms are given against given concentration ranges for potassium-40 and strontium-90, ruthenium-106 and caesium-134, and caesium-137 and cerium-144, respectively. Extremely high values are excluded from the illustrations. While strontium-90 (Fig.1), caesium-134 (Fig.2) and caesium-137 (Fig.3) give relatively sharp peaks in these illustrations, the peaks for potassium-40 (Fig.1), ruthenium-106 (Fig.2) and cerium-144 (Fig.3) are rather flat due to the scatter of the values reported. To measure the three latter radionuclides, the majority of the participating laboratories used non-destructive γ-spectrometry on the dried seaweed. It is striking that considerable scatter was observed for these radionuclides, even though the majority of the results were obtained without employing radiochemical separation procedures. This fact indicates that the calibration of the measuring instruments had not been performed properly in many cases.

Probable concentration values

Table VI gives the probable concentrations in AG-I-1 sample for potassium-40, strontium-90, ruthenium-106, caesium-134, caesium-137

TABLE V. AVERAGE VALUES AND RANGES OF THE REPORTED VALUES FOR THE SEAWEED SAMPLE AG-I-1

Radionuclide	^{40}K	^{60}Co	^{65}Zn	^{90}Sr	^{95}Zr-Nb	^{106}Ru	^{110}Ag^m	^{134}Cs	^{137}Cs	^{144}Ce
No. of reported results	22	20	9	24	23	36	12	31	41	29
Max. value (pCi/g)	79.2	11	25	15.6	50	4800	13.1	715	5047	206
Min. value (pCi/g)	11.3	0.4	0.1	2.4	0.347	28	1.1	6	31.6^a	7.9
Average (pCi/g)	42.2	3.1	4.3	9.6	6.0	287	2.6	34	213	30
σ	± 3.4	± 0.6	± 2.6	± 0.5	± 2.5	± 131	± 1.0	± 24	± 136	± 8
(mean)	(8%)	(19%)	(60%)	(5%)	(42%)	(46%)	(38%)	(71%)	(64%)	(27%)

a The value represents ^{134}Cs + ^{137}Cs.

and cerium-144 (at the reference date of 1 Jan. 1972). The estimations of
these concentrations were made in two ways; first, by applying Chauvenet's
criterion to the groups of data in order to reject outlying data and calculating
the average values of the data lying within the Chauvenet's range; secondly,
by taking high frequency ranges, i.e. the highest frequency range and
neighbouring ranges from Figs 1-3 and calculating the average values of
the data lying within these ranges. The results of the estimations by both
methods are given in Table VI. The Chauvenet's ranges are always wider
than those of the high frequency ranges so that the latter method gives

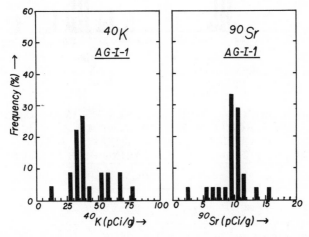

FIG. 1.   Per-cent frequency distribution of the reported results against concentration ranges for potassium-40
and strontium-90 in AG-I-1.

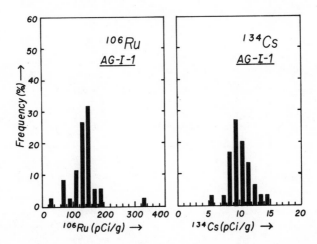

FIG. 2.   Per-cent frequency distribution of the reported results against concentration ranges for ruthenium-106
and caesium-134 in AG-I-1.

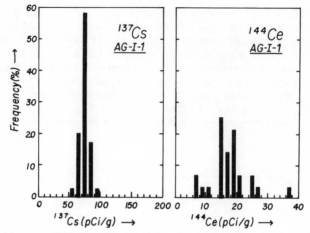

FIG.3. Per-cent frequency distribution of the reported results against concentration ranges for caesium-137 and cerium-144 in AG-I-1.

TABLE VI. PROBABLE CONCENTRATIONS ESTIMATED FOR POTASSIUM-40, STRONTIUM-90, RUTHENIUM-106, CAESIUM-134, CAESIUM-137 AND CERIUM-144 IN AG-I-1 BY APPLYING CHAUVENET'S CRITERION AND BY TAKING HIGH FREQUENCY RANGES

Radionuclide	^{40}K	^{90}Sr	^{106}Ru	^{134}Cs	^{137}Cs	^{144}Ce
No. of reported results	22	24	36	31	41	29
Chauvenet's range (pCi/g)	9 - 72	8.7 - 11.3	70 - 200	5.9 - 14.3	62.2 - 87.4	5.8 - 28.4
No. of values in Chauvenet's range	21	17	32	28	32	26
Average in Chauvenet's range (pCi/g)	40	10.0	135	10.1	75	17.1
σ (mean)	± 3	± 0.1	± 5	± 0.3	± 1	± 0.9
	(7.5%)	(1.0%)	(3.7%)	(3.0%)	(1.3%)	(5.3%)
The range of the highest frequency (pCi/g)	35 - 40	9 - 10	141 - 160	9 - 10	71 - 80	14 - 16
No. of values in high frequency ranges	14	18	22	23	33	17
Average in high frequency ranges (pCi/g)	34.6	10.0	144	10.0	75	17.0
σ (mean)	± 1.2	± 0.2	± 3	± 0.2	± 1	± 0.5
	(3.5%)	(2.0%)	(2.1%)	(2.0%)	(1.3%)	(2.9%)

TABLE VII. PROBABLE CONCENTRATIONS ESTIMATED FOR
COBALT-60, ZINC-65, ZIRCONIUM-NIOBIUM-95 AND SILVER-110m
IN AG-I-1 BY APPLYING CHAUVENET'S CRITERION

Radionuclide	^{60}Co	^{65}Zn	^{95}Zr-Nb	^{110}Ag^m
No. of reported results	20	9	23	12
Chauvenet's range (pCi/g)	1.5 - 2.3	0 - 3.7	0 - 6.1	1.1 - 2.0
No. of values in Chauvenet's range	13	8	16	10
Average in Chauvenet's range (pCi/g)	1.9	1.7	2.9	1.5
σ (mean)	± 0.1	± 0.4	± 0.4	± 0.1
	(5.3%)	(24%)	(14%)	(6.7%)

slightly narrower standard deviations, although the average values estimated
are practically identical. In the case of potassium-40, the estimates by the
two methods differ by about 15% due to a somewhat skewed distribution of
the data in the higher concentration ranges. The estimate by the latter
method may be slightly more probable.

Table VII gives the probable concentrations in AG-I-1 sample for
cobalt-60, zinc-65, zirconium-niobium-95 and silver-110m (at the reference
date of 1 January 1972) estimated by applying Chauvenet's criterion to the
reported data. In these cases the method for estimation by taking high
frequency ranges is not reliable as a result of less data being available.
Since the concentrations of these radionuclides are low, being a few picocuries per gram dried matter, these estimates are less reliable compared
with the six radionuclides mentioned above. The standard deviations of
these estimates range between ±5 and ±24%.

6. ANALYTICAL METHODS ADOPTED

Because of the relatively high radionuclide concentrations in the seaweed
as a sample for monitoring level, and also because of the increased availability
of Ge(Li) detectors for γ-spectrometry, one of the characteristic features
of this intercalibration exercise is the wide use of direct γ-spectrometry
with Ge(Li) detectors on the dried seaweed. Tables VIII-XI list methods for
measurements used by different laboratories for the determinations of
major radionuclides such as strontium-90, ruthenium-106, caesium-134
and 137 and cerium-144. Practically all results for other radionuclides
were obtained by non-destructive γ-spectrometry, except for some cases
of zirconium-niobium-95 and silver-110m. Although it is difficult to
correlate the quality of the results obtained with the methods adopted on the

TABLE VIII. ANALYTICAL METHODS USED BY DIFFERENT
LABORATORIES FOR THE DETERMINATION OF STRONTIUM-90 IN
THE SEAWEED SAMPLE AG-I-1

Method No.	Brief description of method	Code No. of Laboratory
Sr- I	Dry ashing, SrCO$_3$ppt, Sr sep. with fum. HNO$_3$, ^{90}Y milking.	6, 13, 15, 22, 29, 31 39
Sr- II	Dry ashing, Sr oxalate ppt, Sr sep. with fum. HNO$_3$, ^{90}Y milking.	8, 35
Sr- III	Dry (or wet) ashing, Sr phosphate ppt, Sr sep. with fum. HNO$_3$, ^{90}Y milking.	4, 10, 32
Sr- IV	Dry ashing, alkaline fusion or other dis. step, ion-exchange sep., SrCO$_3$ppt, ^{90}Y milking.	9, 20, 24
Sr- V	Dry ashing, alkaline fusion, Sr sep. with fum. HNO$_3$, extr. of ^{90}Y with HTTA-TBP/MIBK, back-extr. in aqu. phase.	45
Sr- VI	Dry (or wet) ashing, extr. of ^{90}Y with HDEHP, stripping in acid, ^{90}Y counting as oxalate or other forms.	11, 14, 30
Sr-VII	Wet ashing, hydroxide scavenge, ^{90}Y milking.	12, 38

TABLE IX. ANALYTICAL METHODS USED BY DIFFERENT
LABORATORIES FOR THE DETERMINATION OF RUTHENIUM-106 IN
THE SEAWEED SAMPLE AG-I-1

Method No.	Brief description of method	Code No. of Laboratory
Ru- I	Direct NaI γ-spec. on dried seaweed.	1, 2, 11, 14, 24, 28 41
Ru- II	Direct Ge(Li) γ-spec. on dried seaweed.	3, 4, 5, 7, 8, 9 10, 13, 15, 16, 19, 20 21, 22, 25, 26, 29, 31 32, 36, 37, 44
Ru-III	Dry ashing, alkaline fusion, RuO$_2$ppt, red. to metallic Ru, β-counting.	33
Ru-IV	Dry ashing, alkaline fusion, extr. of Ru-oxides with CCl$_4$, RuO$_2$ppt, β-counting.	17, 35
Ru- V	Wet ashing, Fe(OH)$_3$ppt, distil. of RuO$_4$, RuO$_2$ppt, β-counting.	12
Ru-VI	Wet ashing, distil. of RuO$_4$, RuO$_2$ppt, red. to metallic Ru, β-counting.	34

TABLE X. ANALYTICAL METHODS USED BY DIFFERENT
LABORATORIES FOR THE DETERMINATIONS OF CAESIUM-134
AND 137 IN THE SEAWEED SAMPLE AG-I-1

Method No.	Brief description of method	Code No. of Laboratory
Cs- I	Direct NaI γ-spec. on dried seaweed.	1, 2, 11, 14, 18, 24 28, 41, 42
Cs- II	Direct Ge(Li) γ-spec. on dried seaweed.	3, 4, 5, 7, 8, 9 10, 13, 16, 19, 20, 21 22, 23, 25, 26, 29, 31 34, 36, 37, 44
Cs- III	Dry ashing, Ge(Li) γ-spec. on the ash.	15
Cs- IV	Dry ashing, AMP abs., (Cs_2PtCl_6ppt), γ-spec. or β-counting.	32, 35, 43
Cs- V	Dry ashing, AMP abs., $Cs_3Co(NO_2)_6ppt$, $Cs_3Bi_3I_9ppt$, Cs_2PtCl_6ppt, β-counting.	6
Cs- VI	Dry ashing, extr. with Phenol-nitrobenzene mix. in dodecyl benzene sulphonate med., Cs_2SnCl_6ppt, β-counting.	30
Cs-VII	Wet ashing, NiFC ppt, AMP abs., Cs_2PtCl_6ppt, β-counting	12

TABLE XI. ANALYTICAL METHODS USED BY DIFFERENT
LABORATORIES FOR THE DETERMINATION OF CERIUM-144 IN THE
SEAWEED SAMPLE AG-I-1

Method No.	Brief description of method	Code No. of Laboratory
Ce- I	Direct NaI γ-spec. on dried seaweed.	1, 2, 11, 14, 24
Ce- II	Direct Ge(Li) γ-spec. on dried seaweed.	3, 4, 5, 7, 8, 9 10, 13, 16, 19, 20, 21 22, 25, 29, 31, 34, 37 44
Ce-III	Dry ashing, Ge(Li) γ-spec. on the ash.	15
Ce-IV	Dry ashing, $Fe(OH)_3ppt$, γ-spec.	35
Ce- V	Wet ashing, $(Fe, La)(OH)_3ppt$, Ce-iodate ppt, β-counting.	12
Ce-VI	Na_2CO_3 fusion, dis. with $HF + HNO_3 + H_2SO_4$ mix., Ce-iodate ppt, β-counting.	36

basis of the results reported by a small number of laboratories which used
identical methods, brief observations on the analytical methods used for the
five major radionuclides are given below.

Strontium-90

    As shown in Table VIII, the radiochemical methods used are categorized
into 7 major schemes. From the reported results obtained by using these
schemes, it seems that most methods can produce good results for the
matrix of this sort at this relatively high activity level. Besides the
standard separation procedure of strontium with fuming nitric acid (Sr-I, -II,
-III) ion exchange separation procedure (Sr-IV) or extraction of yttrium-90
with HDEHP (Sr-VI) seems to work well. The direct yttrium-90 milking
procedure after hydroxide scavenging (Sr-VII) seems to give slightly
deviated results on either the low or high side, although these deviations
might be merely accidental. The extraction of yttrium-90 with the
HTTA-TBP/MIBK system (Sr-V) gave a considerably low result.

Ruthenium-106

    Six different procedures used for ruthenium-106 determination are given
in Table IX. As can be seen from the table, the majority of the laboratories
used γ-spectrometry and only 4 laboratories out of 34 used β-counting. The
major problem in this case lies in the calibration of the γ-spectrometers,
especially those with Ge(Li) detectors. Taking only the results obtained by
Ge(Li) γ-spectrometers the range of the variation is from 28 to 4800 pCi/g.
This fact indicates that, even using sophisticated instruments with high
analytical capability, critical views on the calibration and functioning of
instruments are always essential. It can be said, at the same time, that
the classical β-counting procedures with radiochemical separations can
produce good results with proper performance, as shown in the cases of
methods Ru-IV and Ru-VI. The procedure of wet ashing followed by iron
hydroxide precipitation (Ru-V) seems to produce low results, due perhaps
to incomplete collection of ruthenium by the hydroxide precipitate.

Caesium-134 and 137

    Table X lists seven different methods for the measurements of
caesium-134 and 137. Beta-counting procedures listed are apparently
incapable of differentiating caesium-134 from caesium-137. Nevertheless,
these procedures can be used reliably to measure the sum of caesium 134 +
caesium-137, except for the method Cs-V, which involves several
decontamination steps and gives a fairly low result. Even with this procedure
one can expect to obtain good results with proper control of the radiochemical
yield. The major problem here is also the γ-spectrometer calibration as
has been indicated for ruthenium-106, although scatter of the reported
results for caesium-137 are much smaller than that of ruthenium-106.

Cerium-144

    Since the γ-energy of cerium-144 is relatively soft, Compton scattering
affects considerably the calibration of a γ-spectrometer for this radio-

nuclide. Although the majority of the laboratories used Ge(Li) γ-spectro-
metry, as shown in Table XI, the scatter of the reported results is
relatively large (Fig. 3). Again, the importance of correct γ-spectrometer
calibration is indicated. The methods with radiochemical separation
procedures listed seem to work properly, except for the method Ce-V, which
produced a slightly lower result.

7. GENERAL REMARKS

While surveying the reported results, the following general observations
were made:

(1) In spite of the fact that the instructions had been given to dry the
seaweed at 105-110°C prior to the analysis and to report the results in the
activity on the reference date, these instructions were often ignored.
These are the first basic items any analyst should consider before engaging
in any kind of radionuclide determination.

(2) The concept of the 'significant figures' of reported results in
relation to their estimated errors was completely disregarded in many
instances. This indicates that some analysts lack a critical view of the
accuracy of their measurements.

(3) Possibly due to the sufficient levels of major radionuclides present
in the seaweed sample for measurements, the comparability of the reported
results on strontium-90 and caesium-137 are fairly satisfactory, except
for a few widely deviated results.

(4) Even by using non-destructive γ-spectrometry with Ge(Li) detectors,
variations in reported results on potassium-40, ruthenium-106 and
caesium-144 are fairly large. The emphasis should be made on a proper
control of calibration procedures of counting instruments for these
radionuclides.

(5) In some cases the deviations of the results seem to be caused
simply by mistakes in calculation. Careful control in this aspect is also
required.

(6) Since the seaweed sample AG-I-1 contains higher levels of radionuclides relative to samples for normal environmental monitoring, the
intercalibration on lower level samples may be required to improve
comparability of results in actual monitoring situations.

ACKNOWLEDGEMENTS

The authors wish to express their thanks for the effort and collaboration
rendered by all participants during the execution of the present intercalibration exercise. They are especially grateful to Mr. A. Preston
and Mr. J.W.L. Dutton from the British Fisheries Radiobiological Laboratory
for their effort in the collection and preparation of the seaweed sample,
and Dr. V.T. Bowen from Woods Hole Oceanographic Institution for his
extensive collaboration in the homogeneity tests of the sample. They are
also indebted to Mrs. D. Vas for the technical assistance in the work and
Miss B. Eisenhut for the manuscript preparation.

The International Laboratory of Marine Radioactivity operates under a tri-partite agreement between the International Atomic Energy Agency, the Government of the Principality of Monaco and the Oceanographic Institute at Monaco. Support for the present work is gratefully acknowledged. This programme was also partially supported by UNESCO funds.

REFERENCES

[1] FUKAI, R., BALLESTRA, S., MURRAY, C.N., "Intercalibration of methods for measurements of fission products in seawater samples", Radioactive Contamination of the Marine Environment (Proc. Symp. Seattle, 1972), IAEA, Vienna (1973) 3-27.

[2] BOWEN, V.T., "Analyses of sea-water for strontium and strontium-90", Reference Methods for Marine Radioactivity Studies, Technical Reports Series No. 118, IAEA, Vienna (1970) 93-127.

[3] WORLD HEALTH ORGANIZATION, 1966, Methods of Radiochemical Analysis, FAO/IAEA/WHO, Geneva, p. 84-93.

[4] FUKAI, R., BALLESTRA, S., RAPAIRE, J.L., "A simple application of least-squares fitting to gamma spectrometry of marine environmental samples: The case of caesium radionuclides", Rapid Methods for Measuring Radioactivity in the Environment (Proc. Symp. Neuherberg, 1971), IAEA, Vienna (1971) 301-10.

[5] FUKAI, R., MURRAY, C.N., "Results of plutonium intercalibration in seawater and seaweed samples", this Report.

ANNEX

LIST OF PARTICIPATING INSTITUTIONS

Australia	Australian Atomic Energy Commission, Research Establishment, Sutherland, N.S.W.	C.J. Hardy W.W. Flynn
Belgium	Royal Institute of Natural Sciences of Belgium, Laboratory of Physical Oceanography, Brussels	E. Peeters
Canada	Atomic Energy of Canada Limited, Chalk River Nuclear Laboratories, Environmental Research Branch, Chalk River, Ontario	W.E. Grummitt G. Lahaie
Denmark	Danish Atomic Energy Commission, Research Establishment Risø, Roskilde	A. Aarkrog
Finland	University of Helsinki, Department of Radiochemistry, Helsinki	J.K. Miettinen T. Rahola
	Institute of Radiation Physics, Helsinki	A. Salo
France	Commissariat à l'Energie Atomique, Centre d'Etudes Nucléaires de Grenoble, Service de Protection et des Etudes d'Environnement, Grenoble	S. Haddad S. Descours
	Commissariat à l'Energie Atomique, Centre d'Etudes Nucléaires de Fontenay-aux-Roses, Service de Recherches Toxicologiques et Ecologiques, Fontenay-aux-Roses	L. Jeanmaire
	Service Central de Protection contre les Rayonnements ionisants, Le Vésinet	P. Pellerin M.L. Remy J.P. Noroni
Germany, Federal Republic of	Institut für Strahlentechnologie der Bundesforschungsanstalt für Lebensmittelfrischhaltung, Karlsruhe	E. Fischer R. Schelenz
India	Bhabha Atomic Research Centre, Health Physics Division, Bombay	A.K. Ganguly B. Patel
Israel	TECHNION-Israel Institute of Technology, Haifa	H. Shafrir J. Laichter
Japan	Japan Analytical Chemistry Research Institute, Tokyo	T. Asari
	Tokyo Kyoiku University, Department of Chemistry, Tokyo	N. Ikeda R. Seki H. Kirita

Japan (cont.)	Environmental Pollution Research Center, Ibaraki Prefectural Hygienic Laboratory, Mito	R. Koike S. Morita
	National Institute of Radiological Sciences, Chiba	M. Saiki T. Ueda
	Power Reactor and Nuclear Fuel Development Corporation, Tokai-mura	H. Sato K. Tsutsumi K. Nagasawa
	Tokai Regional Fisheries Research Laboratory, Tokyo	H. Tsuruga T. Umezu Y. Minamisako
	National Institute for Public Health	N. Yamagata K. Iwashima
Korea, Republic of	Atomic Energy Research Institute, Health Physics Division, Seoul	Kyung Rin Yang
New Zealand	Department of Health, National Radiation Laboratory, Christchurch	J.E. Dobbs
Norway	Institutt for Atomenergi, Kjeller	E. Stedje E. Steinnes R. Garder
Poland	Polish Academy of Sciences, Marine Station P.A.N. Sopot	R. Bojanowski
Romania	Polytechnical Institute "Gheorghe Gheorghiu-Dej", Bucharest	I.I. Georgescu
South Africa	Atomic Energy Board, Isotope and Radiation Division, Pretoria	J.K. Basson D. van As
Sweden	Aktiebolaget Atomenergi, Health and Safety Section, Studsvik	P.O. Agnedal G. Bergström H. Tovedal
United Kingdom	British Nuclear Fuels Limited, Windscale Works, Health and Safety Branch, Sellafield, Cumberland	H. Howells T.H. Boyd
	Fisheries Radiobiological Laboratory, Lowestoft, Suffolk	A. Preston N.T. Mitchell J.W.R. Dutton
United States of America	Woods Hole Oceanographic Institution, Woods Hole, Mass.	V.T. Bowen H.D. Livingston
	National Bureau of Standards, Analytical Chemistry Division, Washington, D.C.	L.A. Currie J.C. May

	Oregon State University, Department of Oceanography, Corvallis, Oregon	N. Cutshall
United States of America (cont.)	Environmental Protection Agency, Southeastern Radiological Health Laboratory, Montgomery, Alabama	D. G. Easterly
	New York University, Medical Center, Institute of Environmental Medicine, New York, N. Y.	M. Eisenbud T. Kneip
	Scripps Institution of Oceanography, La Jolla, California	T. R. Folsom V. F. Hodge K. M. Wong
	Puerto Rico Nuclear Center, Mayaguez	W. O. Forster E. D. Wood
	Brookhaven National Laboratory, Health Physics and Safety Division, Upton, New York	A. P. Hull
	National Bureau of Standards, Radioactivity Section, Center for Radiation Research, Washington, D. C.	J. M. R. Hutchinson J. R. Noyce
	New York State, Dept. of Health, Radiological Science Laboratories, Albany, New York	J. M. Matuszek
	Environmental Protection Agency, National Environmental Research Center, Cincinnati, Ohio	D. M. Montgomery B. Kahn
	Atlantic Richfield Company, Richland, Washington	G. C. Oberg H. E. Smith C. H. McLoughlin
	Battelle Memorial Institute, Pacific Northwest Laboratories, Richland, Washington	D. E. Robertson
	University of Washington, Fisheries Centre, Seattle, Washington	W. R. Schell
USSR	Academy of Sciences of UkSSR, Institute of Biology of South Seas, Sevastopol	L. G. Kulebakina V. N. Korableva
	All-Union Research Institute of Marine Fisheries and Oceanography (VNIRO), Moscow	S. A. Patin A. A. Petrov
Yugoslavia	Yugoslav Academy of Sciences and Arts, Institute for Medical Research, Zagreb	M. Sarić
IAEA	Seibersdorf Laboratory	O. Suschny
	Monaco Laboratory	R. Fukai S. Ballestra D. Vas

RADIOACTIVITY IN THE HYDROSPHERE
Recent trends in Czech studies

A. MANSFELD
Water Resources Research Institute,
Prague, Czechoslovakia

Abstract

RADIOACTIVITY IN THE HYDROSPHERE: RECENT TRENDS IN CZECH STUDIES.
Recent trends in the study of radioactivity in the hydrosphere in Czechoslovakia are given. One of the objects of Czechoslovakia's national programme of water conservation is to obtain information on the behaviour of radioactive substances in water courses, reservoirs, sources of drinking water and purification plants. Systematic measurement of radioactivity in surface waters and drinking water began in 1962. The study is directed mainly at measurement of total beta and ^{90}Sr activity, although ^{226}Ra and uranium activities were also determined. Causes of radioactive pollution are assessed, e.g. effluent from uranium mining and processing. The results are used in comparing the quality of effluent with established health safety standards.

Considerable attention is being paid to the radioactivity level of water in cases where radioactive waste is being discharged into water bodies or buried in the ground or where higher concentrations of radioactive substances occur in water owing to the presence of certain types of rock. The main emphasis is on the health hazards to the domestic users of water, geological research and the needs of certain special branches of industry.

Because of certain geological factors, radioactivity in water became a subject of study in Czechoslovakia even before the Second World War, the first measurements being performed on waters with a high mineral content. However, only radon concentrations were studied, i.e. even high concentrations of dissolved matter (up to 45 g/l in water containing ionic bromine) did not pose any particular methodological difficulties. Systematic study of the radioactivity in surface waters and drinking water began in 1962.

The study programme has had a number of different aims and its orientation has been (and is being) changed considerably so that the results always provide answers to practical questions connected with the prevention of water pollution. The first aim was the study of the influence of radioactive fall-out at the time of the nuclear weapons tests. Sampling of river water was carried out at 120 points; at the same time 130 major sources of drinking water were investigated. The study was directed mainly at the determination of total beta and ^{90}Sr activity; ^{226}Ra and uranium activities were also measured. It was found that, although the radioactivity level of water samples rose, the increase was only temporary and the maximum permissible level was exceeded at only a few points. At that time, and in 1963, only 30% of the four million people covered by the study were drinking water with a total beta activity level of less than 5 pCi/l; by 1969 the figure was virtually 100%. The drop in the ^{90}Sr concentration during the subsequent period can be seen from, for example, average data compiled at the Homutov waterworks (surface water, coagulation, filtration). The ^{90}Sr concentration, which was 1.3 pCi/l in 1964, had fallen to 0.8 pCi/l by 1965 and 0.5 pCi/l by 1969.

Attention was then focussed on causes of radioactive pollution of water (for example, effluent from uranium mining and processing). Workers determined quantities of effluent, the specific activity of individual radionuclides and the chemical composition of effluent. The results were used in comparing the quality of effluent for which health safety standards had been established and in calculating the material balances in effluent on the basis of selected radionuclides.

Study of the radioactivity of the hydrosphere is now part of Czechoslovakia's national programme of water conservation. The objective of this programme, which is based on 75-80 sampling points and an annual sampling frequency of 12-24, is to obtain information on the behaviour of radioactive substances in watercourses, reservoirs, sources of drinking water and purification plants. The main emphasis is on determining ^{226}Ra, natural uranium, ^{90}Sr and total beta activity; there are, however, also methods for determining ^{210}Pb, ^{210}Po, natural thorium, ^{137}Cs and total alpha activity.

The reason for orienting the programme in this way is that we have only an approximate idea of how radioactive waste products behave in water bodies, as their behaviour is influenced by a large number of physical, chemical and biological processes, which complicate the picture of radioactivity migration formed under straightforward laboratory conditions and make it difficult to assess the threat to man and his environment. Owing to lack of information, many purification plants have had to be built without a thorough study of the interrelationship between the radioactivity source, health safety requirements, the water body in question and the size of the purification plant, and without the establishment of technical and economic optima. The above-mentioned complex interrelationship has usually been represented in oversimplified form as a balance between the quantity of radioactive material and the flow rate in the water body, the gradual interaction between radioactive substances and hydrosphere being ignored.

By now, considerable experimental data have been accumulated on the distribution of different radionuclides between the water, biomass and bottom sediments. By means of correlation and regression analysis it has been possible to establish for several radionuclides empirical relationships between the coefficients of accumulation in freshwater organisms. As the information about the behaviour of radionuclides in different water body components as a function of changing environmental factors becomes more precise, we think that mathematical models should be developed for studying the processes whereby radionuclides migrate in the hydrosphere. Mathematical models can be very important in solving a number of practical problems — especially that of predicting the fate of radionuclides in water bodies.

Factors that limit both the extent and the quality of work on the behaviour of radionuclides in the hydrosphere are the reproducibility and applicability of the results of radiochemical analyses. Methods of determination have, therefore, been gradually modernized and are being supplemented to meet the needs of research and day-to-day operation. The following review indicates the present state of the art.

RADIUM-226

Emanometric determination

Water samples are sealed in a receptacle and radon allowed to accumulate. After a certain time the accumulated radon is transferred by means of a circulation system to the ionization chamber of an electrometer where the radioactivity is measured. The instrument is calibrated using an emanating solution with a known ^{226}Ra concentration.

Determination by co-precipitation

The radium is separated out by co-precipitation with lead and barium sulphates. The sulphates are dissolved in an alkaline solution of EDTA. The barium (radium) sulphate is precipitated from the resulting solution by glacial acetic acid. The precipitate is mixed with a luminophore. A sorption-emulsion technique is used for carrying out the determination.

The method is based essentially on the selective adsorption of radium ions. The radium is adsorbed onto glass plates covered with a layer of gelatine containing barium sulphate. The measurements are performed with an alpha scintillation unit. The activity of the solution is calculated using the internal standard method.

LEAD-210

Determination by the extraction method

This method consists essentially in the formation of lead dithizonate dissolved in chloroform. The lead dithizonate is extracted in cyanide at pH 8-9. The lead is re-extracted into the aqueous phase by dilute nitric acid and precipitated in chromate form. The ^{210}Pb activity is determined by measuring the equilibrium amount of ^{210}Bi.

Determination using an ion exchange unit

In the determination of lead advantage is taken of this element's ability to become adsorbed on to an anionite from $1\underline{N}$ HCl. The ^{210}Pb is easily removed from the anionite by water, the daughter isotope — ^{210}Bi — remaining on the anionite. The ^{210}Bi can be removed from the anionite by $2\underline{N}$ H_2SO_4.

POLONIUM-210

The method consists essentially in the preferential adsorption of polonium onto a luminophore.

URANIUM

Determination by fluorimetry

Exposure of uranium in solidified melts to ultra-violet light with a wavelength of about 366 μm gives rise to fluorescence with a maximum in the range 550-565 μm, which is proportional to the uranium concentration. Either 98% NaF + 2% LiF or 9% NaF + 45.5% Na_2CO_3 + 45.5% K_2CO_3 is used as the melt. The determination is carried out either visually or by means of a fluorimeter.

Photocolorimetric determination using 'arsenazo III'

The uranium is separated out by sorption on silica gel from a solution containing ammonium chloride, tartaric acid and EDTA. The uranium is separated from the silica gel by acetic acid and determined photocolorimetrically with the help of 'arsenazo III'.

THORIUM

A sulphoacid cationite is used in the preliminary concentration and separation of thorium, which is removed by a saturated solution of ammonium oxalate and determined photocolorimetrically with the help of 'arsenazo III'.

STRONTIUM-90

The strontium is precipitated in carbonate form, the precipitate dissolved in hydrochloric acid and the strontium and calcium oxalates precipitated out of the solution. Barium is separated out in chromate form. The precipitate is dissolved and, after the addition of an yttrium salt carrier, is left for ^{90}Y to form. The separation of the yttrium is based on the precipitation of the hydroxide, which is then transferred to the oxalate for measurement.

CAESIUM-137

The preliminary concentration of the caesium is achieved by co-precipitation with iron ferrocyanide; the caesium is extracted in the form of dipicryl amine in nitrobenzene.

From the point of view of physical dosimetry, these methods are based on the detection of beta or alpha radiation. With appropriate instruments and detectors, the lowest measurable concentration for a measuring time of 30 minutes is about 0.8 pCi/l in the case of ^{210}Pb, 0.2 pCi/l in the case of ^{90}Sr, etc.

The methods have a wide range of applications, including the analysis of surface, underground and waste waters with different chemical compositions. In investigations relating to existing and future mineral water springs, carried out in 1971, certain difficulties were encountered in the case of waters with dissolved matter concentrations of 20-50 g/l. By slightly modifying the analytical procedures, however, it was possible to overcome these difficulties. The methods described above for determining different radionuclides in water can, if certain modifications are made, also be used for bottom sediment and biomass investigations.

HOW TO ORDER IAEA PUBLICATIONS

Exclusive sales agents for IAEA publications, to whom all orders and inquiries should be addressed, have been appointed in the following countries:

UNITED KINGDOM	Her Majesty's Stationery Office, P.O. Box 569, London SE 1 9NH
UNITED STATES OF AMERICA	UNIPUB, Inc., P.O. Box 433, Murray Hill Station, New York, N.Y. 10016

In the following countries IAEA publications may be purchased from the sales agents or booksellers listed or through your major local booksellers. Payment can be made in local currency or with UNESCO coupons.

ARGENTINA	Comisión Nacional de Energía Atómica, Avenida del Libertador 8250, Buenos Aires
AUSTRALIA	Hunter Publications, 58 A Gipps Street, Collingwood, Victoria 3066
BELGIUM	Service du Courrier de l'UNESCO, 112, Rue du Trône, B-1050 Brussels
CANADA	Information Canada, 171 Slater Street, Ottawa, Ont. K 1 A 0S 9
C.S.S.R.	S.N.T.L., Spálená 51, CS-11000 Prague
	Alfa, Publishers, Hurbanovo námestie 6, CS-80000 Bratislava
FRANCE	Office International de Documentation et Librairie, 48, rue Gay-Lussac, F-75005 Paris
HUNGARY	Kultura, Hungarian Trading Company for Books and Newspapers, P.O. Box 149, H-1011 Budapest 62
INDIA	Oxford Book and Stationery Comp., 17, Park Street, Calcutta 16
ISRAEL	Heiliger and Co., 3, Nathan Strauss Str., Jerusalem
ITALY	Libreria Scientifica, Dott. de Biasio Lucio "aeiou", Via Meravigli 16, I-20123 Milan
JAPAN	Maruzen Company, Ltd., P.O.Box 5050, 100-31 Tokyo International
NETHERLANDS	Marinus Nijhoff N.V., Lange Voorhout 9-11, P.O. Box 269, The Hague
PAKISTAN	Mirza Book Agency, 65, The Mall, P.O.Box 729, Lahore-3
POLAND	Ars Polona, Centrala Handlu Zagranicznego, Krakowskie Przedmiescie 7, Warsaw
ROMANIA	Cartimex, 3-5 13 Decembrie Street, P.O.Box 134-135, Bucarest
SOUTH AFRICA	Van Schaik's Bookstore, P.O.Box 724, Pretoria
	Universitas Books (Pty) Ltd., P.O.Box 1557, Pretoria
SPAIN	Nautrônica, S.A., Pérez Ayuso 16, Madrid-2
SWEDEN	C.E. Fritzes Kungl. Hovbokhandel, Fredsgatan 2, S-10307 Stockholm
U.S.S.R.	Mezhdunarodnaya Kniga, Smolenskaya-Sennaya 32-34, Moscow G-200
YUGOSLAVIA	Jugoslovenska Knjiga, Terazije 27, YU-11000 Belgrade

Orders from countries where sales agents have not yet been appointed and requests for information should be addressed directly to:

Publishing Section,
International Atomic Energy Agency,
Kärntner Ring 11, P.O.Box 590, A-1011 Vienna, Austria

75-04822